Functionally Graded Materials (FGMs)

Functionally Graded Materials (FGMs)

Fabrication, Properties, Applications, and Advancements

Edited by
Pulak M. Pandey, Sandeep Rathee, Manu Srivastava,
and Prashant K. Jain

CRC Press
Taylor & Francis Group
Boca Raton London New York

CRC Press is an imprint of the
Taylor & Francis Group, an **informa** business

First edition published 2022
by CRC Press
6000 Broken Sound Parkway NW, Suite 300, Boca Raton, FL 33487-2742

and by CRC Press
2 Park Square, Milton Park, Abingdon, Oxon, OX14 4RN

© 2022 Taylor & Francis Group, LLC

CRC Press is an imprint of Taylor & Francis Group, LLC

Library of Congress Cataloging-in-Publication Data

Names: Pandey, Pulak M., editor.
Title: Functionally graded materials (FGMs) : fabrication, properties, applications, and advancements / edited by Pulak M. Pandey, Sandeep Rathee, manu Srivastava, and Prashant K. Jain.
Description: Boca Raton : CRC Press, 2022. | Includes bibliographical references and index.
Identifiers: LCCN 2021013063 (print) | LCCN 2021013064 (ebook) | ISBN 9780367483814 (hbk) | ISBN 9780367564858 (pbk) | ISBN 9781003097976 (ebk)
Subjects: LCSH: Functionally gradient materials.
Classification: LCC TA418.9.F85 F88 2022 (print) | LCC TA418.9.F85 (ebook) | DDC 620.1/18--dc23
LC record available at https://lccn.loc.gov/2021013063
LC ebook record available at https://lccn.loc.gov/2021013064

ISBN: 978-0-367-48381-4 (hbk)
ISBN: 978-0-367-56485-8 (pbk)
ISBN: 978-1-003-09797-6 (ebk)

DOI: 10.1201/9781003097976

Typeset in Times
by Deanta Global Publishing Services, Chennai, India

Contents

Preface

The domain of functionally graded materials (FGMs) is enormous and attracts the keen interest of researchers worldwide. Corresponding to enhanced understanding of the concepts of material customization, the global manufacturing scenario has witnessed remarkably enhanced FGM applications. The fabrication of FGMs has also seen numerous innovative trends over the last few decades. The aim of this book is to strengthen the collaborative research practices undertaken by FGM research groups worldwide and to present a consolidated idea of different aspects of FGMs to the materials and manufacturing researchers.

The editors of this book started working together as a research team in different aspects of materials and realized the severe need to consolidate the undergoing research work on different aspects of FGM worldwide. Effort has been undertaken by the editing team to include various aspects of FGMs. The chapters included in this book have been briefly introduced here to make the reader well versed with the overall content.

Chapter 1 discusses different FGMs concepts, especially their types, fabrication methods, and application areas. The chapter concludes with a discussion on various ongoing research efforts in the field of FGM.

Chapter 2 details the different FGM fabrication techniques including liquid phase, vapor phase, and solid phase processing. It proceeds to discuss FGM deposition techniques and also overview different additive manufacturing (AM) techniques for FGM fabrication in practice. It concludes with a discussion on challenges and future potential in the domain of FGM fabrication.

Chapter 3 describes various liquid phase processing techniques for FGMs. The chapter presents a detailed discussion on centrifugal casting, centrifugal slurry pouring, centrifugal pressurized casting, slip casting, tape casting, infiltration, and gel casting methods. The chapter concludes with a critical analysis of the various liquid phase processing techniques of FGM fabrication.

Chapter 4 details various gaseous phase processing techniques for FGMs such as thermal spray, physical vapor, and chemical vapor deposition techniques. It further presents an overview of the computational analysis, and mechanical and thermal behavior of FGMs especially with respect to gaseous phase techniques. It concludes with a summary of applications of FGMs especially in nanocomposites, thermal barrier coatings, bioceramics, concrete structures, etc.

Chapter 5 presents a detailed discussion on different AM methods for FGM fabrication. This includes directed energy deposition, powder bed fusion, extrusion-based, vat photopolymerization, material jetting, and hybrid AM methods. This is followed by a discussion on the state-of-the-art material systems. The chapter finally concludes with a brief summary of challenges, future potential, and prospects in the domain of FGM fabrication via the AM route.

Chapter 6 presents a case study for the biomedical application of FGM fabricated via the fused filament fabrication method. Study of human bone density variation at various cross-sections using digital image processing technique is undertaken.

An algorithm capable of extracting useful data from an image based on the density variation of bone and its variability which is dependent on the value of grayscale for each pixel is developed. The chapter concludes with a discussion of the utility of FGMs in biomedical applications.

Chapter 7 discusses the recent advancement in the analysis of FGM structures. Their bending, vibration, and buckling analysis is then discussed in detail. The chapter concludes with a detailed discussion on the future scope of FGM structures.

Chapter 8 presents a case study involving the analysis and modeling of smart FGM structures. A discussion on their bending and vibration aspects is then presented. Smart composite materials and structures are discussed. Nonlinear static analysis of smart FG plates combined with active fiber composite as piezoelectric material is then presented. Simulation models are developed and validated for these plates using AYNSYS software. The chapter concludes with a summary of the case study.

Chapter 9 presents a case study on the dynamic analysis of porous sandwich functionally graded material plate with geometric nonlinearity. Sandwich plate with a homogeneous core has been analyzed for the nonlinear thermodynamic response. The geometrical nonlinearity in the von Karman sense is considered.

Chapter 10 overviews different applications, recent trends, and the future scope of FGMs.

This book is completed with an intent to provide anyone a one-stop destination for understanding all the aspects of the FGMs starting from their conceptual understanding to their fabrication methods, simulation, and modeling techniques to case studies, application to trends, etc. Its aim is to be a landmark work for understanding, fabricating, analyzing, and applying different FGMs by various research groups, including manufacturing and material science practitioners. The primary motivation behind the present book is to fulfil the societal responsibility of sharing knowledge with fellow researchers and the manufacturing community and to create a platform from which altogether newer avenues may emerge. The editors have tried their level best to include different aspects of FGMs. The contributors have specifically been requested to keep the language at its simplest best technically to enable readers from all over the world to utilize the information shared in this work. The editor group most humbly welcomes all the queries, advice, and observations from readers towards improvement of the quality of our work in future book editions. The editor group also prays to the Almighty that this work serves its purpose of benefiting its readers and enabling them to come up with their own innovations in the field of FGMs.

MATLAB® is a registered trademark of The MathWorks, Inc. For product information, please contact:

The MathWorks, Inc.
3 Apple Hill Drive
Natick, MA 01760-2098 USA
Tel: 508 647 7000
Fax: 508-647-7001
E-mail: info@mathworks.com
Web: www.mathworks.com

Acknowledgments

Words fall short in expressing our affection and gratitude for the editor Mrs. Cindy Renee Carelli for agreeing to publish this work and for her immense support and cooperation as well as coordination. She is inarguably one of the best professionals who is a pleasure to work with. Her intense and honest feedback coupled with the ability to add a human touch to everything can enhance the quality of any project she is a part of. A word of appreciation is also due to Ms. Erin Harris for all her timely support. The editors thank the entire team of CRC Press for their support towards the initiative of bringing forth this edited book.

This being an edited book, it would not have come into being without the indispensable contribution and support of all our authors who devoted a lot of energy in enabling us to bring this book into its current form. We express deep gratitude to them for their valuable contribution.

The editor Dr. Pulak M. Pandey wishes to use this platform to acknowledge his supervisors and teachers. The editor devotes his professional success to the deep values of righteousness and morality imbibed from his parents Shri A. N. Pandey and Smt. Rati Pandey. The invaluable support of his siblings, well-wishers, and students is also deeply acknowledged. No words of admiration and affection suffice for the understanding and support of his wife Smt. Kiran Pandey and his loving children Harsh and Mudit. And last but not least he is thankful to his editorial team members and various authors for the unmatched support he received for the chapter writing and book editing.

The editor Dr. Sandeep Rathee wishes to thank his mentor Dr. Pulak M. Pandey for placing trust in and supporting his abilities. The editor expresses immense gratitude towards his director Prof. Rakesh Sehgal. The support received from Prof. Nazir Ahmad Sheikh shall always be cherished. The editor conveys heartfelt thanks to his supervisors Dr. Sachin Maheshwari and Dr. Arshad Noor Siddiquee. He also acknowledges the support of his family, especially his parents Shri Raj Singh Rathee and Smt. Krishna Rathee, Dr. Manu, Mr. Sombir Rathee, Mrs. Manju, and Mrs. Anita for their constant support.

The editor Dr. Manu Srivastava wishes to thank her director Dr. Sanjeev Jain for placing immense trust in her capabilities. She also thanks her institute officials, faculty members, and her students for their inspiration and support. The author wishes to convey her heartfelt thanks to her supervisors Dr. T. K. Kundra and Dr. Sachin Maheshwari. No words of gratitude would suffice to thank her mentor in the field of additive manufacturing and functionally graded materials Dr. Pulak M. Pandey. She takes this platform to acknowledge the support of her friends and family, especially her late parents, her son Yash Krishna Srivastava, Mr. Ram Krishna Yashaswi, Dr. Sandeep Rathee, Dr. T. K. Singh, Ms. Rachna Johar, and Ms. Shilpi Chauhan, for their affectionate support and unfaltering trust.

The editor Dr. Prashant K. Jain wishes to thank his Director Dr. Sanjeev Jain for all his support. He thanks and acknowledges with affection the support of his colleagues, students, and staff members. The editor thanks his mentor and supervisor Dr. Pulak M. Pandey. He acknowledges with a deep sense of gratitude the support of his parents Shri M. L. Jain and Smt. Asha Jain, his brothers Vijayant Kumar Jain and Rashmi Kant Jain, his uncles Shri D. K. Jain and Shri N. K. Jain, his wife Mrs. Nidhi Jain, and his beloved sons Pal and Prateek.

The editors most humbly dedicate this edited book to the Divine Creator based upon the belief that the strength to bring any thoughtful and noble endeavor into being emanates from the Almighty. The editor group honestly hopes that the present book makes a valuable addition in the field of FGMs.

Editors

Pulak M. Pandey is presently serving the Indian Institute of Technology Delhi as a Professor in the Department of Mechanical Engineering. His previous teaching assignment was at Harcourt Butler Technical University Kanpur, India. He was awarded his Ph.D. degree from the Indian Institute of Technology Kanpur. His areas of interest mainly include additive manufacturing/3D printing and tooling, CAD/CAM, nontraditional machining and finishing, FEA of manufacturing processes, and the biomedical application of 3D printing. He is a recipient of numerous awards including but not limited to five Gandhiyan Young Technological Innovation Awards, an Outstanding Research Award by the Additive Manufacturing Society of India, a J M Mahajan Award of outstanding faculty by the Dept. of Mech. Engg., IIT Delhi, an Outstanding Young Faculty Fellowship (IIT Delhi) sponsored by the Kusuma Trust, Gibraltar, a Highly Commended Paper Award by *Rapid Prototyping Journal* for the paper "Experimental investigations into the effect of delay time on part strength in Selective Laser Sintering", a Certificate of Merit for achieving first position in B. Tech. Mechanical Engineering in Kanpur University, a Merit Certificate for proficiency in physics, short listed for an INAE Young Engineer award, numerous expert talks at international and national levels, etc. Dr. Pandey is on the editorial committees for many high-indexed journals. Dr. Pandey has around 250 publications in various international platforms of repute.

Sandeep Rathee is currently serving the Department of Mechanical Engineering, National Institute of Technology Srinagar, India, as an Assistant Professor. His previous assignment was as a Post-Doctoral Fellow at Indian Institute of Technology Delhi (IIT Delhi). He is the recipient of the prestigious National Post-Doctoral Fellowship by SERB (Govt. of India). Prior to this, he worked with Amity School of Engineering and Technology, Amity University Madhya Pradesh, India. He was awarded Ph.D. degree from Faculty of Technology, University of Delhi. His field of research mainly includes friction stir welding/processing, advanced materials, composites, additive manufacturing, advanced manufacturing processes, and characterization. He has authored over 60 publications in various international journals of repute and refereed international conferences. He has authored three books on "Additive manufacturing: fundamentals and advancements"; "friction based solid state additive manufacturing techniques" and "Functionally Graded Materials: Fabrication, Properties, Applications, and Advancements" with CRC press, Taylor & Francis group. He has a total teaching and research experience of around ten years. He has delivered invited lectures, chaired scientific sessions in several national and international conferences, STTPs, and QIP programs. He is a life member of the Additive Manufacturing Society of India (AMSI), and Vignana Bharti (VIBHA).

Manu Srivastava is presently serving the PDPM Indian Institute of Information Technology, Design and Manufacturing Jabalpur, India, in the Department of Mechanical Engineering as an Assistant Professor. Her previous assignment was as a Head, Department of Mechanical Engineering and Director Research, Faculty of Engineering and Technology, Manav Rachna International Institute of Research and Studies, Faridabad, India. She completed her Ph.D. in the field of additive manufacturing from Faculty of Technology, University of Delhi. Her field of research is additive manufacturing, friction-based AM, friction stir processing, advanced materials, manufacturing practices, and optimization techniques. She has around 70 publications in various technical platforms of repute. She has authored two books on additive manufacturing: *Fundamentals and Advancements*; and *Friction Based Solid State Additive Manufacturing Techniques* with CRC Press. She has a total teaching and research experience of around 15 years. She has won several proficiency awards during the course of her career including merit awards, best teacher awards, proficiency awards, etc. She has delivered invited lectures, chaired scientific sessions in several national and international conferences, STTPs, and QIP programs. She is a life member of the Additive Manufacturing Society of India, Vignana Bharti (VIBHA), the Institution of Engineers (IEI India), the Indian Society for Technical Education (ISTE), the Indian Society of Theoretical and Applied Mechanics (ISTAM), and the Indian Institute of Forging (IIF).

Prashant K. Jain is Associate Professor and Dean (Students) at PDPM Indian Institute of Information Technology, Design and Manufacturing Jabalpur, India. He has over 25 years of research and teaching experience at UG and PG levels. He graduated in mechanical engineering from Dr. H. S. Gour University Sagar in 1995 and did a master's degree in advanced production systems at Samrat Ashok Technological Institute (S.A.T.I.) Vidisha. He received a Ph.D. degree in the area of rapid prototyping/additive manufacturing, particularly on part strength studies in selective laser sintering processes from the Indian Institute of Technology Delhi (IIT Delhi) in 2009.

Dr. Prashant K. Jain was previously serving the Delhi Technological University (formerly known as Delhi College of Engineering) as a lecturer in the Mechanical Engineering Department. He has also served the Indian Institute of Technology Delhi as a project scientist and research associate where he was engaged in research work on error compensation in CNC machines and part quality improvement in fused deposition modeling.

He has more than 110 publications to his credit, published in international peer-reviewed journals of repute, and national and international conferences in India and abroad. He has guided 5 Ph.D. and 27 master's students, with several ongoing. He has been a resource person and delivered invited lectures, chaired scientific sessions in several national and international conferences, STTPs, and QIP programs. His research interests extend from geometric modeling, CAD/CAM integration, incremental sheet forming, computational geometry, rapid prototyping and tooling, and incremental sheet forming to nano technologies in manufacturing.

List of Contributors

E. T. Akinlabi
Pan Africa University for Life and
    Earth Sciences Institute
Ibadan, Nigeria

and

Department of Mechanical Engineering
    Science
University of Johannesburg
Johannesburg, Republic of South Africa

Stephen Akinlabi
Department of Mechanical Engineering
Walter Sisulu University
Butterworth, Republic of South Africa

S.T. Aruna
Surface Engineering Division
CSIR-National Aerospace Laboratories
Bengaluru, India

Pushkal Badoniya
Department of Mechanical Engg.
PDPM Indian Institute of Information
    Technology
Jabalpur, India

H. D. Chalak
Department of Civil Engineering
National Institute of Technology
Kurukshetra, India

Brijesh Gangil
Department of Mechanical Engineering
H.N.B.U.
Garhwal, India

Aman Garg
Department of Civil Engineering
National Institute of Technology
Kurukshetra, India

Vivek Kumar Gupta
FFF Lab, Mechanical Engineering
    Discipline
PDPM Indian Institute of Information
    Technology
Jabalpur, India

Suraj Prakash Harsha
Vibration and Noise Control Laboratory
Indian Institute of Technology
Roorkee, India

Sunir Hassan
Department of Mechanical Engineering
Walter Sisulu University
Butterworth, Republic of South Africa

Prashant K. Jain
FFF Lab, Mechanical Engineering
    Discipline
PDPM Indian Institute of Information
    Technology
Jabalpur, India

T. C. Jen
Department of Mechanical Engineering
    Science
University of Johannesburg
Johannesburg, Republic of South Africa

Sharnappa Joladarashi
Department of Mechanical Engineering
National Institute of Technology
   Karnataka
Surathkal, India

Praveennath G. Koppad
Department of Mechanical Engineering
National Institute of Technology
   Karnataka
Surathkal, India

Rasheedat Mahamood
Department of Material and
   Metallurgical Engineering
University of Ilorin
Ilorin, Nigeria

and

Department of Mechanical Engineering
   Science
University of Johannesburg
Johannesburg, Republic of South
   Africa.

Evgenii Murashkin
Ishlinsky Institute of Problems in
   Mechanics
Russian Academy of Science
Moscow, Russian Federation

Ankit Nayak
FFF Lab, Mechanical Engineering
   Discipline
PDPM Indian Institute of Information
   Technology
Jabalpur, India

and

School of Automation
Banasthali Vidyapith
Rajasthan, India

Rityuj Singh Parihar
Department of Mechanical Engineering
National Institute of Technology Raipur
Chhattisgarh, India

M. R. Ramesh
Department of Mechanical Engineering
National Institute of Technology
   Karnataka
Surathkal, India

Lalit Ranakoti
Department of Mechanical Engineering
National Institute of Technology
Uttrakhand, India

Sandeep Rathee
Department of Mechanical Engg.
National Institute of Technology
Srinagar, Jammu & Kashmir, India

Nagaraja C. Reddy
Department of Mechanical Engineering
Bangalore Institute of Technology
Bengaluru, India

Raj Kumar Sahu
Department of Mechanical Engineering
National Institute of Technology Raipur
Chhattisgarh, India

Saroj Kumar Sarangi
Department of Mechanical
   Engineering,
National Institute of Technology
Patna, India

Srinivasu Gangi Setti
Department of Mechanical Engineering
National Institute of Technology Raipur
Chhattisgarh, India

Michael Shatalov
Department of Mathematics and
    Statistics
Tshwane University of Technology
Pretoria, Republic of South Africa

C. Siddaraju
Department of Mechanical Engineering
Ramaiah Institute of Technology
Bengaluru, India

Simran Jeet Singh
MPAE Division, Mechanical
    Engineering Department
Netaji Subhas University of Technology
Dwarka, New Delhi, India

Agnivesh Kumar Sinha
Department of Mechanical Engineering
National Institute of Technology Raipur
Chhattisgarh, India

Manu Srivastava
Department of Mechanical Engineering
PDPM Indian Institute of Information
    Technology
Jabalpur, India

T. Varol
Department of Metallurgical and
    Materials Engineering
Karadeniz Technical University
Trabzon, Turkey

Shashikant Verma
Department of Mechanical Engineering
I.E.T, Bundelkhand University
Jhansi, India

Ashish Yadav
Department of Mechanical Engineering
PDPM Indian Institute of Information
    Technology
Jabalpur, India

1 Functionally Graded Materials

An Introduction

Rasheedat Mahamood, TC Jen,
Stephen Akinlabi, Sunir Hassan, Michael Shatalov,
Evgenii Murashkin, and Esther Akinlabi

CONTENTS

1.1   INTRODUCTION

Functionally graded materials (FGM), as the name implies, are materials that are tailored based on the functionality required from the material. Functionally graded materials were first proposed in Japan in the 1980s to be used in a space project (rocket engine) [1]. The researchers were confronted with the problem of developing a material that could withstand a high temperature difference. The traditional composite materials that were tried for this project kept failing due to delamination of the metal and ceramic materials used. The failure was attributed to the existence of a sharp distinct interface between the materials. These researchers thought that if the sharp interface between the materials of the laminate composite could be eliminated, then the composite failure could be averted. That was when the researchers came up with the idea that, instead of joining two dissimilar materials, the materials should be gradually introduced to each other. Traditional laminate composite materials involve joining two different materials where the interface between the two materials is a transition of one material into another in an instance. This interface is a site of high concentration factor and the site where failure is usually initiated.

DOI: 10.1201/9781003097976-1

1

The reason is that the two materials recognize each other as different materials, and it is easy for the two materials to de-bond when subjected to extreme working conditions. Imagine a scenario where a material is gradually added to another material such that the first or the base material only sees the second material as an impurity that is scattered within it. If such a base material is subjected to harsh working conditions, it will be difficult to separate the second material from the first. Figure 1.1 shows a schematic diagram of FGM and traditional composite material. The Japanese researchers discovered that by replacing the sharp distinct interface between two materials in a laminate composite material with a gradual or graded interface, then the problem of delamination in the composite material will be solved. These researchers developed a novel composite material by gradually introducing the second material into the base material such that the second material was gradually increased in percentage in the base material from 0% to 100%. In this way, the sharp interface was eliminated, and a novel material called FGM was developed. Since then, FGMs have found applications in numerous fields [2].

Functionally graded materials exist in nature such as bone, teeth, wood, and bamboo [2, 3]. By examining each of these materials, one finds out why nature made them to be so. Teeth consist of an external part made of very hard material while the core is made up of soft and tough material. These two materials have different properties, and they can perform the function that nature intended of them with their conflicting properties. Enamel is the white and the hardest outer part of a tooth that is made of calcium phosphate. Dentin is a hard tissue consisting of microscopic tubes underlying the enamel. Pulp is softer tissue where blood vessels and nerves run through, underlying the cementum. The cementum is a layer of connective tissue binding the roots of the teeth firmly to the gums and jawbone. Whenever nature wants to add two mismatched materials, there is usually an FGM in the interface or transition zone of such materials. Another example of this is the transition zone between the bone and cartilage. Bone is a hard material while cartilage is a soft material; in between a bone and cartilage is a transition zone that is FGM. Figure 1.2 shows a diagram of a typical human bone [2]. It can be seen how nature has crafted these dissimilar materials to serve the functionality requirement of a tooth. Scientists have always

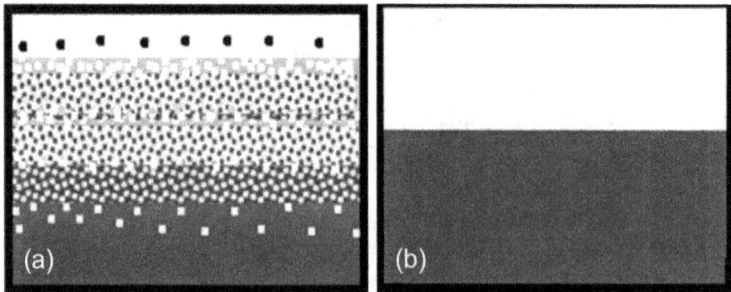

FIGURE 1.1 (a) Schematic diagram of a functionally graded composite material and (b) traditional laminate composite material [2].

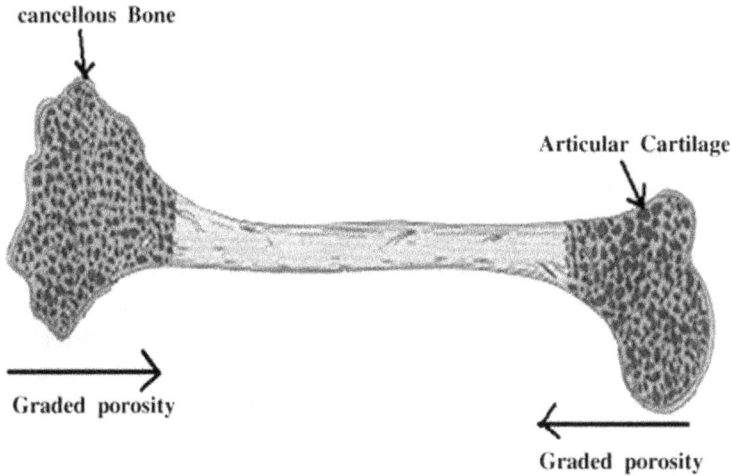

FIGURE 1.2 Schematic diagram of human bone showing FGM structures [2].

taken cues from nature to solve scientific and technological problems. Artificial neural networks are an example of that. An artificial neural network imitates the human brain neural network and how it processes information. Artificial neural networks have been used to solve many engineering problems. Functionally graded materials have also been used to solve several engineering problems that range from the field of automobiles to structural elements [4]. In this chapter, an overview of different types of FGM is presented. The methods of fabrication and areas of application are also highlighted. Some of the research efforts in this field are also presented.

1.2 FUNCTIONALLY GRADED MATERIALS IN NATURE

Nature has a way of teaching us lessons. Scientists and engineers have always turned to nature when they are confronted with one problem or another. Functionally graded materials exist in nature serving their intended functions very well which has inspired scientists and engineers. Bamboo is an example of naturally existing FGM with a non-uniform distribution of constituent material and of varying microstructure [2]. Wood is another important FGM existing in nature. Wood is made up of cellulose and lignin with the lignin forming the matrix while the cellulose is dispersed in varying degrees across the entire material. Some tissues and organs in the human body are naturally occurring FGMs, including human skin, bone, and teeth. Whenever there is attachment of one tissue to another in the human anatomy such as in joints, the transition is usually through FGMs. Such transition zones are such that one material is gradually introduced to the other material until the new material completely replaces the initial material. By so doing, the stress concentration that is usually associated with having two different materials joined is eliminated. Examples of these include the transition between bone and cartilage, bone and tendon, etc. These transition zones are called ligaments [2]. There are quite a lot of FGMs existing in nature that have inspired scientists and engineers and have been used to solve engineering and medical problems. There are different types of functionally graded materials that have been created after the first was produced in Japan. The next section discusses the various types of FGMs.

1.3   TYPE OF FUNCTIONALLY GRADED MATERIALS

Functionally graded materials can be broadly classified into three types, namely: porosity gradient, compositionally or chemical gradient, and microstructure gradient FGM [2]. An example of a porosity gradient FGM that occurs in nature is bone; the varying degree of porosity gives bone its sponginess characteristics. A few artificial porosity gradient FGMs have been created for different engineering and biological applications [4, 5]. Compositionally graded materials are other types of FGM that occur naturally; examples include human teeth. Several compositionally graded materials have also been created by scientists for engineering and medical applications. Microstructural gradient FGMs occur naturally in bamboo. Artificial microstructurally gradient FGMs are produced usually in a monolithic material such as pure metal. They are achieved by solidifying a molten metal in a controlled manner, that is, by varying the cooling rate, eventually yielding the desired controlled varying microstructure [2].

1.4   METHODS OF FABRICATION OF FUNCTIONALLY GRADED MATERIALS

Functionally graded materials are used as surface coatings (thin coatings) and as bulk material depending on the intended application. Functionally graded thin coatings are usually produced when there is a need for the surface properties to be different from the properties of the bulk materials. For example, if the surface of a material will be engaging in a sliding motion with another surface, this makes it necessary for such material to have a surface property with high wear resistance. If the bulk material does not possess such property, then it becomes imperative to fabricate a hard surface with wear resistance properties. A functionally graded thin coating is the choice here because the sliding wear mechanism will subject the material to harsh working conditions which will result in rapid breakdown if the coating is not functionally graded. A functionally graded thin coating ensures that the coating does not break down when materials are subjected to an extreme working environment. There are various methods now available for manufacturing functionally graded materials [6]. These fabrication methods are classified into two categories, namely: the fabrication of functionally graded thin coatings and the fabrication of functionally graded bulk materials [2]. Methods of fabrication of functionally graded thin coatings include the chemical vapor deposition process, physical vapor deposition process, friction stir processing, self-propagating high-temperature synthesis process, and additive manufacturing processes. Methods that have been used to fabricate bulk functionally graded materials include powder metallurgy, casting, and additive manufacturing processes. Each of these fabrication methods has been used to successfully fabricate functionally graded thin coatings or functionally graded bulk materials [7]. All the fabrication processes for functionally graded thin coatings mentioned except additive manufacturing are employed after the part has been produced as a finishing manufacturing process. Additive manufacturing (AM) processes on the other hand are advanced manufacturing technologies that can produce three-dimensional (3D) components directly from a 3D computer-aided model of the part simply by adding materials layer after layer until the fabrication is completed [8]. In addition, the manufacturing method can also be used to fabricate parts made up of FGM as well as parts with thin coatings of FGM. These can be achieved in one manufacturing run. That is, unlike in other processes where the part is first produced with different manufacturing processes and then finished with another process to produce the FGM thin coating, AM can be used to produce parts no matter the complexity and with FGM bulk material or a surface thin FGM coating in one manufacturing run. AM has the capability to use multiple materials and when programmed into the AM machine, the desired part can be produced and with the required compositional make-up and desired surface properties. AM is a layered manufacturing process in which materials are added layer by layer, making it easy to vary the material composition as desired. Some of the areas of application of FGMs are presented in the next section.

1.5 AREAS OF APPLICATION OF FUNCTIONALLY GRADED MATERIALS

Functionally graded materials have found applications in many fields, and the application areas also keep expanding. Some of the areas of application of FGM include the aerospace, biomedical, automobile, defense, electrical, electronics, energy, marine, opto-electronics, and materials development industries. Functionally graded materials are now being used to develop new materials because of the unprecedented properties evolving from them. With the development of solid-state material processing techniques, novel materials can now be developed. The combination of dissimilar materials that cannot be achieved using traditional manufacturing technologies is now possible with advanced manufacturing technologies such as additive manufacturing technologies and friction stir processing leading to the development of novel high-performance functionally graded bulk and thin coating materials [9–16]. Some application areas of FGM are summarized in Figure 1.3 [2].

1.6 RESEARCH EFFORTS IN FUNCTIONALLY GRADED MATERIALS

Functionally graded materials have gained a lot of attention in the research community due to the importance of these revolutionary materials [17–20]. There has been constant striving to improve the FGM material properties and in the development of novel FGM materials through combinations of dissimilar materials. Some of these research works are highlighted in this section.

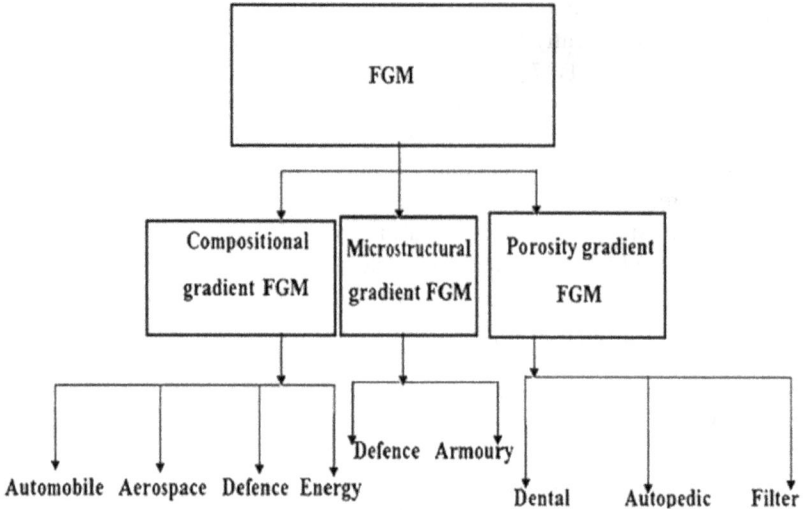

FIGURE 1.3   Areas of application of FGM [10].

Santhosh et al. [21] developed a numerical model for FGM using friction stir processing and studied the mechanical properties and temperature distributions with COMSOL Multiphysics software. The authors carried out a numerical analysis of the developed aluminum 6061 T6 alloy and SiC in-situ metal matrix using friction stir processing. The temperature distribution of the developed FGM material was analyzed with a three-dimensional finite element method using COMSOL Multiphysics software. The study revealed that linear property variation exists over the entire volume of the produced functionally graded materials. Rajasekhar et al. [22] developed four dissimilar-layered volume fractions of FGMs of Al–Cu using powder metallurgy. Premixed powders at different volume fractions of Cu powder (0, 5, 10, and 15 wt.%) were hot pressed using a pressure of 400 MPa and then sintered at 550°C for 3 h each layer at a time and in turn. Studying the mechanical and microstructural properties of the FGM shows that the material composition distribution changes gradually across the layers. The study revealed an improvement in the mechanical properties of the FGM of Al–Cu produced. Also, the densification was found to increase with increasing Cu content in the FGM. Chandrasekaran et al. [23] investigated the suitability and effectiveness of FGMs for marine risers. The authors proposed a high corrosion-resistant functionally graded material of duplex stainless steel and carbon-manganese steel for the risers for marine applications using wire-arc additive manufacturing (WAAM). The investigation revealed that there is a marginal increase in the ultimate strength of produced FGM which was attributed to the formation of a fine solidified microstructure without any defects. The diffusion of elements was also observed to have occurred at the interface causing a coalescence zone between the two materials. Secondary phases are also seen in the microstructure consisting of pearlite, bainite, and martensite finely distributed in the ferrite matrix. This is responsible for the variation of hardness values observed along the build direction. The corrosion resistance behavior was found to be greatly improved. Shen et al. [24] produced an FGM of Fe–Fe$_3$Ni using a wire-arc additive manufacturing process. The authors studied the as-fabricated FGM and heat-treated FGM. The as-fabricated FGM was found to contain both fcc-Fe$_3$Ni and bcc-a-Fe while the heat-treated FGM samples contained mainly bcc-a-Fe that had dissolved into the Fe$_3$Ni matrix that resulted in lower hardness in the heat-treated FGM. The microstructural study showed that the heat-treated FGM sample contained considerable growth of columnar grains in the region with high Ni content and transformed the cellular structures seen in the low-Ni-content region into bainite and residual austenite. Mohebbi et al. [25] investigated the use of numerical methods based on the finite-difference method to solve steady-state heat conduction problems in FGMs. Two different types of material gradations, namely, quadratic and exponential material gradations, were investigated. The results revealed that the proposed numerical approach is robust, and the thermal conductivity was not much affected by the introduction of a significant measurement error. Markandan et al. [26] investigated the feasibility of fabricating FGMs using selective heat melting (SHM). The authors successfully produced a 3D-printed iron (Fe)–polyethylene (PE) composite, a copper (Cu)–polyethylene (PE) composite, and functionally graded CuO foams using the SHM technique. The study revealed that to produce fully dense materials, a low feed rate, high airflow rate, and high airflow temperature were required to efficiently deliver heat from the emitted hot air to the powder bed so that the PE binder particles can melt and form dense composites with smooth surfaces. This study showed that selective heat melting can be used to produce multi-material composites and FGMs. Figure 1.4 shows scanning electron microscopy images of an FGM of CuO foams with the Cu fraction ranging from 10 vol.% to 30 vol.% Cu.

Su et al. [27] fabricated Inconel-steel FGMs using additive manufacturing technology by varying the composition of Inconel in 316 L incrementally from 0% to 100% and studied the influence of the composition gradient on the microstructure and mechanical properties. The results showed that the microstructure of all the FGM samples consisted of columnar dendrites and fine equiaxed grains. It was observed that when the Inconel718 content exceeded 40%, laves phase precipitates were seen in the interdendritic regions. Grain coarsening was also observed in FGM samples containing a relatively small composition gradient of between 5% and 10% due to serious heat accumulation. The micrographs of the FGM samples are shown in Figure 1.5.

The microhardness was found to first decrease at low percent composition and then increase. As the composition gradient decreased, the range of microhardness variation increased along the deposition direction. The fracture mechanism of 316/Inconel718 FGMs was due to aggregate of microporous which was caused by laves phase precipitation thereby forming the micro-pores. The highest ultimate tensile strength of 527.05 Mpa and the highest elongation of 26.21% were attained by a sample containing a composition gradient of 10%. In another study conducted by Luginina et al. [28], spark plasma sintering (SPS) was used to produce an FGM of HA/bioactive glass. The study revealed that the optimized process parameters of 800°C/2 min/50 MPa produced a highly dense FGM sample with better mechanical

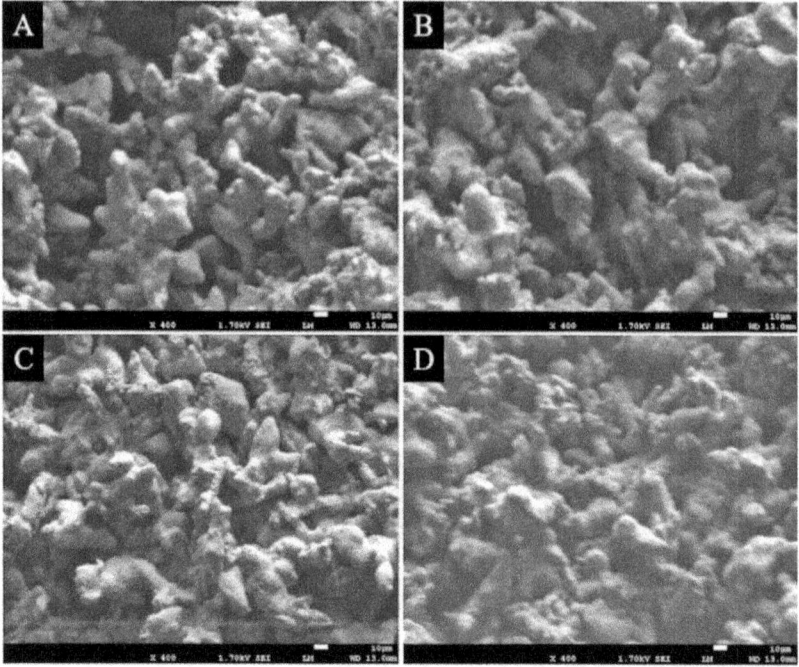

FIGURE 1.4 Scanning electron microscopy images of CuO foams with Cu fraction of (A) 10 vol.%, (B) 15 vol.%, (C) 20 vol.%, and (D) 30 vol.% [26].

and biological characteristics in simulated body fluid (SBF). The authors concluded that the FGM produced can be tailored for biological response for specific clinical applications. Pradeep and Rameshkumar [29] reviewed the production of functionally graded materials using a centrifugal casting method. The influence of processing parameters such as processing temperature, cooling rate, rotational speed of mold, particle size, and density on the properties of fabricated centrifugally cast FGM was analyzed. A proper understanding of processing parameters in the centrifugal casting method is useful in the designing of FGMs. Geng et al. [30] studied the fabrication of Ti6Al4V FGM using laser surface modification (LSM). A microstructural FGM was produced on the surface of Ti6Al4V in this study. The surface of the Ti6Al4V was modified to consist of an FGM microstructure which was achieved by applying nine sets of different LSM process parameters to these samples. The results showed three microstructural zones consisting of a martensitic layer, a transition layer, and the unprocessed or original microstructure of the Ti6Al4V. The transition zone microstructure was seen beneath the martensitic zone which is on the surface of the sample consisting of larger, equiaxed grains and some martensitic α phase, i.e., martensitic α, equiaxed α, and the grain boundary β. The authors also developed a dual phase crystal plasticity finite element model to predict the tensile property of the FGM. The hardened surface zone prevented the propagation of continuous slip bands, while the transition zone prevented excessively sharp stress concentrations

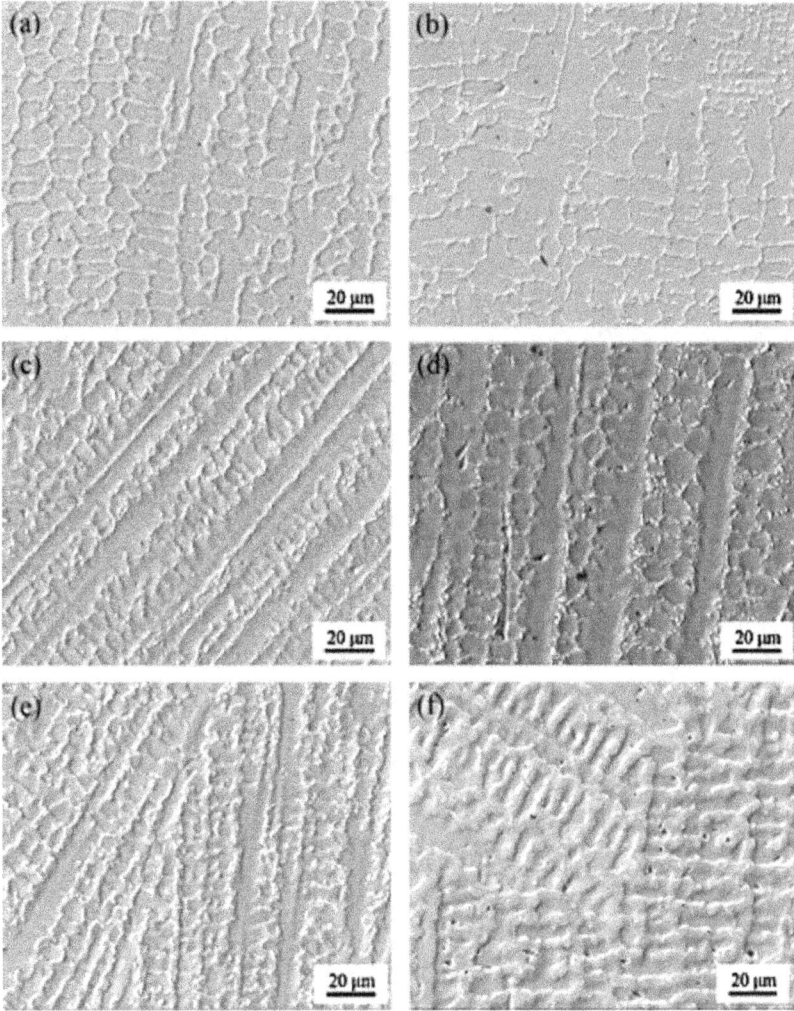

FIGURE 1.5 The microstructure of 316 L/Inconel718 FGM obtained from component A: (a) 100% 316 L; (b) 80% 316 L + 20% Inconel718; (c) 60% 316 L + 40% Inconel718; (d) 40% 316 L + 60% Inconel718; (e) 20% 316 L + 80% Inconel718; (f) 100% Inconel718 [27].

between the outer surface and interior of the samples. The results revealed that the hardened martensitic alpha surface zone helped to improve the tensile property by preventing the propagation of continuous slip bands. Also, the transition zone helps to reduce stress concentrations and prevent the breakdown potential of the martensitic surface and the equiaxed microstructure at the core of the samples, that is, the original microstructure.

An overview of research progress in metallic FGMs produced using laser metal deposition processes was conducted by Yan et al. [31]. Recent findings in the design, fabrication, and characterization of different types of metallic FGMs using LMD were analyzed. Also, the challenges in producing metallic FGMs through the LMD route were highlighted. Model development and numerical simulation were also proposed for further studies. Zhang et al. [32] produced fully dense FGMs of TiC–Ni using combustion synthesis and the hot compaction of Ti, C, and Ni powders under a hydrostatic pressure simultaneously. Finite element analysis was employed to optimize the composition gradient achieved by stacking different powder mixtures of desired compositions. The results revealed that the microstructure varies gradually throughout the volume of the samples. The mechanical properties of the produced FGM were found to be dependent on the amount of Ni content with the optimum at an Ni content of 20 wt.%. The maximum in fracture toughness was obtained at 30 wt.% Ni. Young's modulus was found to decrease with increasing Ni content.

1.7 SUMMARY

FGMs have been introduced in this chapter. Japanese researchers were confronted with the problem of composite failure through delamination when trying to develop a suitable composite material for a harsh working environment in a space project. This quest to solve the problem led to the development of FGM. FGMs exist in nature, and some of them are presented. Types of FGMs and areas of application of FGMs are also highlighted. Functionally graded materials are an active research area, and some of the current research works in this field are also presented.

ACKNOWLEDGMENTS

This project was supported by NRF-IRG SA/Russia Bilateral and technical research collaboration grant No.: NRF/UID No. 118905 and NRF/RFBR No. 19-51-60001.

REFERENCES

1. Hirai, T., Chen, L. Recent and prospective development of functionally graded materials in Japan. *Materials Science Forum*, 1999. 308–311: 509–514.
2. Mahamood, R.M., Akinlabi, E.T. Introduction to functionally graded materials. In: *Functionally Graded Materials. Topics in Mining, Metallurgy and Materials Engineering*, 2017. Springer, Cham. https://0-doi-org.ujlink.uj.ac.za/10.1007/978-3-319-53756-6_1
3. Mahamood, R.M., Akinlabi, E.T, *Functionally Graded Materials*, 2017. Springer, Cham.
4. Naebe, M., Shirvanimoghaddam, K. Functionally graded materials: A review of fabrication and properties, *Applied Materials Today*, 2016. 5: 223–245.
5. Mahamood, R.M., Akinlabi, E.T., Shukla, M., Pityana, S. Characterizing the effect of processing parameters on the porosity of laser deposited titanium alloy powder. In: Proceedings of the International MultiConference of Engineers and Computer Scientists (IMECS, Hong Kong). March 12–14, 2014.
6. Gasik, M.M. Functionally graded materials: Bulk processing techniques. *International Journal of Materials and Product Technology*, 2010. 39(1/2): 20–29.

7. Parihar, R.S., Setti, S.G., Sahu, R.K. Recent advances in the manufacturing processes of functionally graded materials: A review. *Science and Engineering of Composite Materials*, 2018. 25(2): 309–336.

8. Mahamood, R.M. *Laser Metal Deposition Process of Metals, Alloys, and Composite Materials*, 2018. Springer, Cham.

9. Zhao, K., Zhang, G., Ma, G., Shen, C., Wu, D. Microstructure and mechanical properties of titanium alloy/zirconia functionally graded materials prepared by laser additive manufacturing. *Journal of Manufacturing Processes*, 2020. 56(Part A): 616–622.

10. Mahamood R.M., Akinlabi E.T. Types of functionally graded materials and their areas of application. In: *Functionally Graded Materials. Topics in Mining, Metallurgy and Materials Engineering*, 2017. Springer, Cham. https://0-doi-org.ujlink.uj.ac.za/10.1007/978-3-319-53756-6_2

11. Liu, Y., Wang, Y., Wu, X., Shi, J. Nonequilibrium thermodynamic calculation and experimental investigation of an additively manufactured functionally graded material. *Journal of Alloys and Compounds*, 2020. 838: 155322.

12. Gandra, J., Miranda, R., Vilaça, P., Velhinho, A., Teixeira, J.P. Functionally graded materials produced by friction stir processing. *Journal of Materials Processing Technology*, 2011. 211(11): 1659–1668.

13. Sharma, A., Bandari, V., Ito, K., Kohama, K., Ramji, M., Himasekhar S.B.V. A new process for design and manufacture of tailor-made functionally graded composites through friction stir additive manufacturing. *Journal of Manufacturing Processes*, 2017. 26: 122–130.

14. Bikkina, K., Talasila, S.R., Adepu, K. Characterization of aluminum based functionally graded composites developed via friction stir processing. *Transactions of Nonferrous Metals Society of China*, 2020. 30(7): 1743–1755.

15. Mahamood, R.M., Akinlabi, E.T. Laser metal deposition of functionally graded Ti6Al4V/TiC. *Materials & Design*, 2015. 84: 402–410.

16. Mahamood, R.M., Akinlabi, E.T., Laser-assisted additive fabrication of micro-sized coatings. In: *Advances in Laser Materials Processing*, in Woodhead Publishing Series in Welding and Other Joining Technologies (Second Edition), J. Lawrence (ed), Woodhead Publishing, 2018: 635–664.

17. Ezzin, D., Wang, B., Qian, Z. Propagation behavior of ultrasonic love waves in functionally graded piezoelectric-piezomagnetic materials with exponential variation. *Mechanics of Materials*, 2020. 148: 103492.

18. Silva, E.C.N., Walters, M.C., Paulino, G.H. Modeling bamboo as a functionally graded material: Lessons for the analysis of affordable materials. *Journal of Materials Science*, 2006. 41: 6991–7004.

19. Sadowski, T., Boniecki, M., Librant, Z., Nakonieczny, K. Theoretical prediction and experimental verification of temperature distribution in FGM cylindrical plates subjected to thermal shock. *International Journal of Heat and Mass Transfer*, 2007. 50: 4461–4467.

20. Muller, E., Drasar, C., Schilz, J., Kaysser, W.A. Functionally graded materials for sensor and energy applications. *Materials Science and Engineering: Part A*, 2003. 362: 17–39.

21. Santhosh, V., Prakash, D.N.A., Murugan, K., Babu, N. Thermo-mechanical analysis of Tailor-made functionally graded materials through Friction stir processing. *Materials Today: Proceedings*, 2020, Article in Press.

22. Rajasekhar, K., Babu, V.S., Davidson, M.J. Microstructural and mechanical properties of Al-Cu functionally graded materials fabricated by powder metallurgy method. *Materials Today: Proceedings*, 2020.

23. Chandrasekaran, S. Hari, S., Amirthalingam, M. Wire arc additive manufacturing of functionally graded material for marine risers. *Materials Science and Engineering: Part A*, 2020. 792.
24. Shen, C., Hua, X. Reid, M. Liss, K.D., Mou, G., Pan, Z., Huang, Y., Li, H. Thermal induced phase evolution of Fe–Fe3Ni functionally graded material fabricated using the wire-arc additive manufacturing process: An in-situ neutron diffraction study, *Journal of Alloys and Compounds*, 2020. 826.
25. Mohebbi, F., Evans, B., Rabczuk, T. Solving direct and inverse heat conduction problems in functionally graded materials using an accurate and robust numerical method. *International Journal of Thermal Sciences*, 2021. 159.
26. Markandan, K., Lim, R., Kanaujia, P.K., Seetoh, I., Rosdi, M.R.M., Tey, Z.H., Goh, J.S., Lam, Y.C., Lai, C. Additive manufacturing of composite materials and functionally graded structures using selective heat melting technique. *Journal of Materials Science & Technology*, 2020. 47.
27. Su, Y. Chen, B., Tan, C., Song, X., Feng, J. Influence of composition gradient variation on the microstructure and mechanical properties of 316L/Inconel718 functionally graded material fabricated by laser additive manufacturing, *Journal of Materials Processing Technology*, 2020. 283.
28. Marina Luginina, Damiano Angioni, Selena Montinaro, Roberto Orrù, Giacomo Cao, Rachele Sergi, Devis Bellucci, Valeria Cannillo, Hydroxyapatite/bioactive glass functionally graded materials (FGM) for bone tissue engineering, *Journal of the European Ceramic Society*, 2020. 40(13).
29. Pradeep, A.D., Rameshkumar, T. Review on centrifugal casting of functionally graded materials. *Materials Today: Proceedings*, 2020.
30. Geng, Y., McCarthy, E., Brabazon, D., Harrison, N. Ti6Al4V functionally graded material via high power and high speed laser surface modification. *Surface and Coatings Technology*, 2020. 398.
31. Yan, L., Chen, Y., Liou, F. Additive manufacturing of functionally graded metallic materials using laser metal deposition. *Additive Manufacturing*, 2020. 31.
32. Zhang, X.H., Han, J.C., Du, S.Y., Wood, J.V. Microstructure and mechanical properties of TiC-Ni functionally graded materials by simultaneous combustion synthesis and compaction. *Journal of Materials Science*, 2000. 3(5): 1925–1930.

2 Advances in Fabrication Techniques of Functionally Graded Materials

Rityuj Singh Parihar, Raj Kumar Sahu, and Srinivasu Gangi Setti

CONTENTS

DOI: 10.1201/9781003097976-2

2.1 INTRODUCTION

Functionally graded materials (FGMs) are combinations of different materials, wherein the microstructure/composition/porosity changes in a continuous or discrete manner along specific dimensions. The material gradient was first designed and created to decrease thermal stresses due to the temperature of ceramic and metal interfaces in 1983 meant for Japan's space shuttle project. The function and design of FGMs can be different externally and internally as per the desired structure and composition. The non-uniform distribution of structure and composition results in the desired properties for specific applications. The idea of composition is not a modern invention; several examples of FGM structure are available in nature, such as culms of barley and bamboo, seashells, bone, and teeth. These structures entirely match modern FGMs in respect of desired functions and properties at the required region. FGMs have numerous applications in the field of engineering such as automotive, aviation, biomedical, the manufacturing and mining industry, etc. [1]. The diverse applications have contributed appreciably to the extensive research on a wide variety of projects related to the development of FGMs. However, the production of FGMs is critical to the successful utilization of the concept of graded material structures.

Thus numerous methods of producing FGMs have emerged over the years to fulfill the demand for these marvelous materials in diverse engineering applications. A range of FGM production methods, for instance chemical vapor deposition, powder metallurgy, physical vapor deposition, and centrifugal casting, etc., are available [2]. Also the state-of-the-art additive manufacturing process has evolved over the years for developing FGMs with a variety of thicknesses (coating or bulk) and material combinations (metal/ceramic/polymers) [2]. From the first reported FGMs to the most recent one, the different production methods can be categorized in several ways. Initially these methods were grouped according to the material combination, such as ceramic/ceramic, metal/ceramic, metal/metal, and ceramic/polymer. A number of categorizations exist to illustrate the FGM production methods by graded type (microstructure, composition, and porosity), state (solid, liquid, and vapor), structure (continuous and discrete), dimensions and scale (thin and thick), graded nature (constructive and transport), feasible form complexity, the field of application, energy consumption, environmental impact, and overall process expenditure [3].

This chapter enumerates the advancement of FGM fabrication techniques over time on the basis of existing research on preparation methods. Here the fabrication techniques are categorized according to the state of materials while processing, so the function and area of application can be recognized to control as well as enhance FGM properties. The operating principles of fabrication techniques along with examples of adopted material systems are explained in brief with advantages and limitations. Initially the liquid phase processing techniques (such as centrifugal casting, tape casting, infiltration, and cast-decant-cast) are discussed, followed by an explanation of vapor phase processing practices (for example physical vapor deposition and chemical vapor deposition, etc.). Subsequently solid phase processing techniques such as powder metallurgy are elaborated. Finally a state-of-the-art additive manufacturing process is described. At the end of the chapter the future potential and conclusions are presented.

2.2 CLASSIFICATION OF FUNCTIONALLY GRADED MATERIAL FABRICATION TECHNIQUES

Efficient fabrication techniques are critical to the commercialization of FGMs. The processing methods available for the preparation of homogeneous structures (for instance centrifugal casting and powder metallurgy) are useful for attaining structures with properties gradient, but the limitations of these methods restrict their generalized application. However, based on the state of material while processing, the existing fabrication techniques can be categorized as liquid phase processing (centrifugal casting, tape casting, infiltration technique, and cast-decant-cast technique), vapor phase processing (physical vapor deposition and chemical vapor deposition), solid phase processing (powder metallurgy techniques, friction stir processing), additive manufacturing, and deposition techniques. These prominent techniques are explained in the following sections briefly.

2.3 LIQUID PHASE PROCESSING TECHNIQUES

There are numerous techniques that can be grouped in the liquid state method of developing FGMs such as tape casting, centrifugal force methods, and infiltration, etc. These processing methods can fabricate materials with graded properties, and the majority of these processes are comparatively less costly. However, the deficiencies of these techniques are the complexity in organizing the gradient and absolute wettability amid materials. Also the metal in molten state can create several defects in the final product. The most effective liquid phase processing techniques (the centrifugal force-based technique, tape casting, infiltration technique, and cast-decant-cast process) adopted by researchers for the preparation of FGMs are explained in the following subsections.

2.3.1 CENTRIFUGAL FORCE-BASED TECHNIQUE

In some of the processing techniques centrifugal force is the driving element in the fabrication of bulk FGMs having a continuous gradient with specific attributes that can be suitable for various industrial uses. With this technique, the centrifugal force generated by the rotation of the mold causes the non-uniform distribution of different material phases in the radial direction. There are a number of processing methods employed in the production of FGMs working on the principle of centrifugal force, for instance centrifugal casting, centrifugal pressurization, and the centrifugal slurry method. Also lots of research has been performed to prepare FGMs from a variety of material systems using centrifugal force as the driving element, which is explained in the following subsections.

2.3.1.1 Centrifugal Casting Technique

Centrifugal casting is a liquid phase processing technique, and it is most viable as well as reliable way to fabricate continuous graded FGMs. In history the first patent was granted in 1809 in the name of A. G. Echardt for using a centrifugal casting technique. However, the commercial use of centrifugal casting was started in Baltimore for the fabrication of pipes in 1848. Subsequently, in 1990, this technique was adopted to create components with a composition gradient, and nowadays it is the most appropriate process applied for the preparation of FGMs, wherein the molten metal is filled via gravity force into the mold under the influence of various operating parameters, i.e. feed speed, rotational speed, and the temperature of mold preheat. The gravity force as well as centrifugal force works upon the mold because of the spinning or rotational motion. The centrifugal force is generated by density variation amid the materials mixed together, after the properly mixed molten metal is poured into the rotating mold. The mixer contains two types of materials; one is the base metal or matrix and the other is reinforcement. The presence of such forces results in the effective filling of the mold with the benefits of successful control of the microstructure and improved mechanical properties [4]. The key benefit of centrifugal casting for the fabrication of FGMs is the ability to develop a continuous gradient by using the centrifugal force. But this method is capable of developing primarily symmetrical and simple shapes such as cylinders, tubes, and bushes, etc., which is the major drawback. Also the gradient is restricted by density difference and the centrifugal force of the ingredient materials.

Abundant research work is available on the preparation of FGMs by centrifugal casting using several matrix materials (aluminum (Al), copper (Cu), magnesium (Mg), etc.) and reinforcement (silicon carbide (SiC), zirconium trialuminide (Al_3Zr), aluminum oxide (Al_2O_3), Si_3N_4, TiC, ZrO_2, etc.). In order to control the desired quality of gradient, several factors need to be considered, i.e. the type of matrix as well as reinforcement material, pouring temperature, rotational speed, weight fraction, mold preheat, and solidification rate. The centrifugal casting of FGM is performed in two kinds of machines, a horizontal type employed in the production of long-piece and a vertical type employed for the fabrication of short-length parts. Figure 2.1 shows a representative diagram of centrifugal casting. The centrifugal casting process for the fabrication of FGMs can be categorized as a centrifugal in-situ system

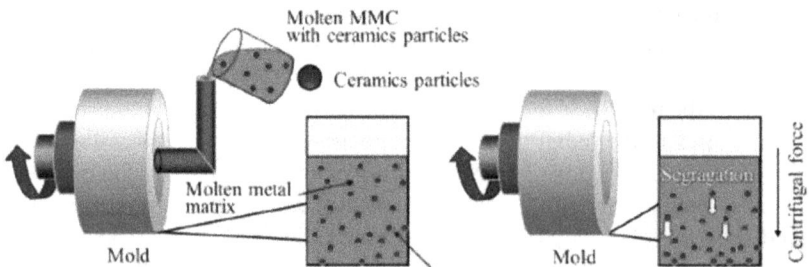

FIGURE 2.1   Representative diagram of centrifugal casting [4].

and centrifugal solid particle system on the basis of the relation between the reinforcement and the matrix.

2.3.1.1.1 Centrifugal Solid Particle Technique

In the centrifugal solid particle technique, a functionally graded tube with gradual property variation is fabricated from a molten metal containing two components, i.e. the matrix and reinforcement. The difference between the melting temperature of the matrix (base metal) and that of the reinforcement (particles) material is used as the driving element in the creation of the composition gradient, wherein the melting temperature of the reinforcement is higher than that of the matrix. Thus the reinforcement remains in solid state while processing and is segregated in the radial direction due to the centrifugal force. The heavier material is collected at the outer and lighter material on the inner region of a functionally graded (FG) tube. The solid particles rotate due to the effect of centrifugal force according to Stoke's law, i.e. larger particles migrate through larger distances.

The commonly adopted matrix materials are Al, Cu, and Mg along with reinforcement materials such as Al_2O_3 and SiC. The quality of the microstructure and the material properties are significantly affected by the mold rotation speed, particle weight fraction, particle size, viscosity of the matrix material, solidification time, and cooling rate. In general the adopted range of rotational speed is 600–2500 revolutions per minute (rpm), and up to 1500 rpm the material properties are improved with satisfactory particle distribution and gradient thickness. Then beyond this rotational speed the increase in centrifugal forces leads to particle agglomeration and property degradation. It was ascertained from earlier research that the revolution speed 1200 rpm is optimum to achieve the graded attributes with satisfactory particle distribution in the thickness/radial direction. Additionally, experiments revealed that increases in particle size and weight fraction also enhance material properties but after a certain limit they were degraded. In earlier research, the weight fraction percentage range varied from 2.5 to 20 wt.%. But the most favorable weight fraction for developing functionally graded composites with satisfactory particle filling capacity was identified as 10wt.% particles by prior research work [5].

2.3.1.1.2 Centrifugal in-Situ Technique

In this method the melting temperature of the reinforcement material is lower than that of the matrix and a gradient is created by applying centrifugal force during solidification. This method is found to be identical to the fabrication of in-situ composite by the crystallization technique. The gradient is developed due to a difference in density between the reinforcement and matrix materials, until the material mixture is in liquid state prior to the primary crystal formation. The primary crystals are generated in the matrix material because of the local chemical composition, followed by precipitation due to the difference in density. The reinforcement distribution mainly depends upon the shape of the end product, and parts produced are highly dense in the outer region compared to the inner. The quality of prepared gradient is significantly influenced by the mold rotation speed, preheating temperature, cooling rate, and pouring temperature [6]. Researchers have developed in-situ

FGMs of Al–Al$_2$Cu, Al–Mg/AlB$_2$, NbC–high chromium white cast iron, and Al–Mg$_2$Si. Also the most preferred processing mold preheat temperature ranges from 30 to 250°C, wherein the optimum condition is 30°C [7]. The preferred mold coating materials are graphite, fiberfrax, and boron nitride, where graphite is most suitable. Moreover the material feed speed should vary from 5 to 50 mm/s, wherein 16 mm/s is optimum for horizontal centrifugal casting machines. The molten material stirring speed and time should vary in the range of 100–600 rpm and 5–20 min, respectively. However the optimum stirring speed and time are 150 rpm and 10 min respectively. Also the reinforcement size ranges from 9 to 120 μm and the ideal size is 16–23 μm. The prime benefit of this method is the ability to develop a continuous gradient, and the limitation is that only cylindrical-shaped FGMs can be fabricated. Another method using centrifugal force for preparing a composition gradient is the centrifugal slurry method [2].

2.3.1.2   Centrifugal Slurry Pouring Technique

Although powder metallurgy process is very advantageous for producing the FGMs, continuous graded properties cannot be prepared. Researchers created a new method named the centrifugal slurry process to fabricate a continuous gradient by using powder metallurgy and centrifugal casting together. In this method two types of particles are present, low-velocity or small-sized or low-density particles and high-velocity or large-sized or high-density particles. These two different kinds of particles are mixed in liquid inside the rotating mold and due to their different characteristic amounts of centrifugal force, the migration rate is also different, which results in a continuous composition gradient. Subsequently liquid slurry is taken out after the completion of sedimentation and green body undergoes a sintering process to get the finished product. However, another difficulty occurred during fabrication due to accumulated high-density particles at the bottom of the rotating mold prior to the completion of the particles spreading throughout the thickness. Therefore it is difficult to create a continuous properties gradient over the whole depth (i.e. a continuous properties gradient can be achieved over 25–35% of the entire depth) [8]. In order to overcome this limitation, Watanabe et al. recommended an innovative approach to creating FGMs with a continuous property gradient over the complete thickness, named the centrifugal slurry pouring method wherein, prior to adding the particles into the liquid, a solvent zone of 100 mm is ensured inside the rotating mold, and finally the full dispersion of particles throughout the thickness without sedimentation was achieved. Another method named centrifugal pressurization was also proposed, wherein the application of centrifugal force generated pressure for the creation of a gradient.

2.3.1.3   Centrifugal Pressurization Methods

In this method the desired continuous gradient is generated prior to the application of centrifugal force. Therefore, the centrifuge method is adopted to apply simple pressure in this process. The centrifugal pressurization methods can be performed in three ways; the first way is the mixed centrifugal powder technique, the second way is the centrifugal sintered-casting technique, and the third way is the reactive centrifugal casting technique [9].

2.3.1.3.1   Mixed Centrifugal Powder Technique

This method is most suited when particle sizes are very small, as the particle distribution in the property gradient mainly depends on the difference of material densities in the mixture. Thus it is very difficult to obtain graded products using particles of small size because the motion of solid particles present in the metallic melt follows Stoke's law, and according to this law a reduction in particle size leads to a decline in the movement velocity. In order to overcome this limitation, Watanabe et al. [10] recommended an innovative way to fabricate FGMs using small-size particles, named the centrifugal mixed power technique, wherein the gradients required for the preparation of nano-scale FGMs should be achieved prior to applying the centrifugal forces. Initially particles of reinforcement metal A and matrix metal B are mixed together and filled in the rotating mold. Subsequently matrix metal B in molten form is poured in the rotating mold over the prior deposited metal A and B particle mixture. The metal B in molten form occupies the inter-particle space of metal A and B. After that molten metal melts the particles of the matrix metal and finally FGMs with enhanced continuous graded properties are obtained. Researchers have developed $Al–TiO_2$, Cu–SiC, and Mg/Mg_2Si, etc., functionally graded composites by using this technique [11]. The volume fraction and size of particles are the most important factors influencing the material properties of developed FGMs. The modification in mixed centrifugal powder technique is done by involving a sintering process and called the centrifugal sintered-casting technique.

2.3.1.3.2   Centrifugal Sintered-Casting Technique

As in the mixed centrifugal powder technique the composition gradient in material mixture is formed before the pouring of molten metal and the application of centrifugal force, but for certain powders it is difficult to achieve. The proposed modification of this method by combining centrifugal casting and centrifugal sintering together is named the centrifugal sintered-casting method [12]. In this method initially the particle mixture of matrix and reinforcement are kept in the rotating mold and then heated to a certain temperature and then held for a fixed time. The heating and holding consolidate the green composition gradient of the particle mixture, and finally the molten matrix metal is poured in, resulting in the proper dispersion of reinforcement particles and fabrication of the desired gradient. The sintering temperature, holding time, and preheating temperature have a significant effect on the mechanical properties and thickness of the gradient. The researchers have adopted copper/diamond, and Al/diamond material systems to develop FGMs [13]. The centrifugal pressurization method was modified by using the exothermic reaction between elemental powder particles, and this process is known as the reactive centrifugal casting technique.

2.3.1.3.3   Reactive Centrifugal Casting Technique

The reactive centrifugal casting technique is mainly utilized in the fabrication of functionally graded layers over tubes or cylinders, where the chances of damage due to oxidation, corrosion, and rusting are high, particularly in the gas and oil supply industry. This method was proposed by Y. Watanabe et al., wherein the exothermic reaction between elementary materials generates heat and results in the formation of an intermetallic compound of comparatively high melting temperature. The intermetallic compound works as reinforcement and disperses in the graded form over the desired thickness [14]. Y. Watanabe et al. fabricated a nickel-aluminides multilayer FGM over the inside surface of a steel pipe using this technique. Initially nickel (Ni) powder was filled inside the rotating steel pipe followed by the pouring of molten aluminum (Al). Consequently an exothermic reaction took place between Al and Ni and finally generated the Ni–aluminide composite coating [15]. The axis of the pipe (vertical or horizontal) or mold rotation and preheating have a very significant effect on the mechanical properties, microstructure, and corrosion resistance of developed FGMs. Apart from centrifugal casting, tape casting is another liquid phase processing technique and is explained in the following section.

2.3.2 TAPE CASTING

The tape casting technique is applied for fabrication using ceramic materials for electronics (insulated substrates and multilayer capacitors), membranes, knives, structural laminates, and solid oxide fuel cells. In this technique the desired mixture of mineral oxide (ceramic) powders is homogeneously mixed with organic binders and additives. The prepared slurry of ceramic powders is placed in a storage container, wherein at the bottom a passage is offered for the supply of slurry. Beneath this passage a polymer tape (made up of non-binding material, i.e. Teflon) passes continuously and slurry is deposited over it and forms a strip. The tape comprises an organic solvent (ethanol) and additives (polymer binder). The thickness of slurry deposited over polymer tape is controlled by a flat knife edge. Subsequently the tape along with the strip passes through the oven to make it dry and finally wound on spool. The complete process of tape casting is also presented in Figure 2.2. The

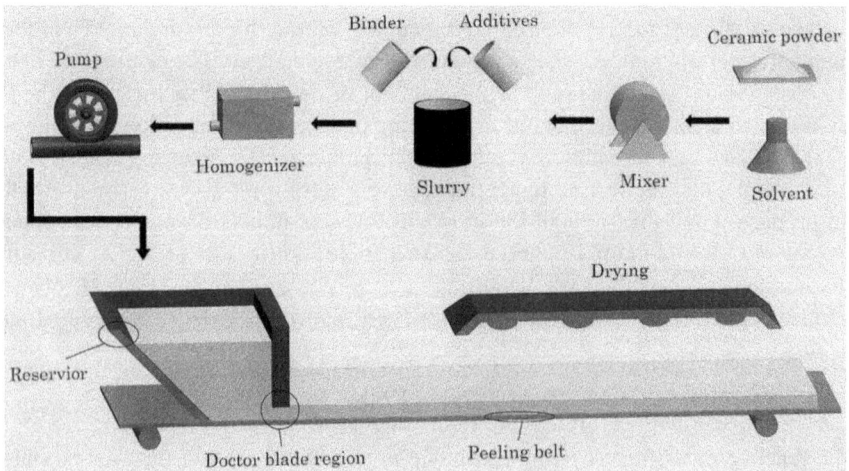

FIGURE 2.2 Representative diagram of tape casting [16].

strips produced by the tape casting technique cover a gradient width from 25 μm to 1000 μm [16]. The tape casting method is combined with the sheet lamination method to obtain a discrete gradient. The strips of different thickness are developed individually and stuck together to form a discrete gradient. Researchers have developed Si–Al–O–N ceramic, W–Cu, and stainless steel 316/ZrO_2 FGMs with 50–1000 μm thickness gradient with improved thermal properties [17]. The infiltration technique is another liquid phase processing technique very suitable for the preparation of ceramic FGMs.

2.3.3 INFILTRATION TECHNIQUE

The infiltration technique can be applied to prepare continuous gradient FGMs, wherein the mold is made up of a porous structure and filled by a slurry of desired materials. In general the mold is made up of gypsum, alumina, plaster of Paris, acrylic, and porous resin, etc. The gypsum mold is most usually employed but it also has deficiencies, for instance the contamination of $CaSO_4$ and lower output due to time elapsed during the drying process. The slurry is filled within the cavities of the porous mold either without applying external pressure (by capillary action) or by means of pressure (mechanical or gaseous action) [18]. After completely filling the mold, the slurry starts reacting by chemical action at the interface and develops a functionally graded structure. Subsequently once the desired thickness of gradient is achieved the slurry is drained out. This method is very effective in preparing a porosity as well as a composition gradient. The commonly adopted material systems for the preparation of FGMs by this method include Cu–W, Al–SiC, and AlN–Cu [19]. Another advanced method based on liquid state processing is the cast-decant-cast process.

2.3.4 CAST-DECANT-CAST TECHNIQUE

This technique was proposed by Scanlan and other materialists of University College Dublin (Ireland) in 2005 [20], for the preparation of wear-resistant, lightweight aluminum–silicon FGMs. This processing method is very useful in the development of FGMs with incompatible material properties (for example: wear resistance-machinability, lightweight-wear resistance, and toughness-hardness) along with cost-efficiency, such as an aluminum cylinder with an iron cylinder liner. Three variants were created for the cast-decant-cast technique with one common step, i.e. decanting: the first material is filled into the mold, then it solidifies over the mold core or wall and sufficient thickness is obtained. Subsequently the remaining unsolidified material is decanted followed by the pouring of another material in the same mold. Three suggested variants of the cast-decant-cast method are turnover, low-pressure, and internal decanting. Although it is a low-cost processing method, it is mainly used for low-melting-temperature material systems and the obtained gradient thickness is also less. Although liquid phase processing techniques have proved useful for the preparation of FGMs, vapor phase processes are also prominent for the preparation of functionally graded coatings.

2.4 VAPOR PHASE PROCESSING TECHNIQUES

Vapor phase processing is the most important technique for developing graded layers of lower depth ranges within nm to sub-mm. In this processing route the desired material in vapor form is condensed over a substrate surface. In general this processing route is used to fabricate functionally graded coating and is categorized as chemical vapor deposition and physical vapor deposition.

2.4.1 PHYSICAL VAPOR DEPOSITION TECHNIQUES

Physical vapor deposition (PVD) is differentiated by a process wherein material transfers to a vapor state from a condensed stage and then returns to a thin condensed layer state. PVD is accomplished by the vaporization of the material and then moved to the substrate as a coating. The PVD is substituted for the electroplating method with comparatively superior material properties. The PVD method is a key process in developing thin FGMs because of its advantages such as the ability to prepare graded thin films with a great range of compounds from metal to alloy deposition and it is comparatively more environment friendly. Typically, at the start of deposition the material chosen for deposition is in solid state, followed by gradual vaporization (atomization) until completely consumed and utilized in the coating process. The working of the PVD process is illustrated in Figure 2.3. This technique has been mainly used in the automotive, aerospace, biomedical, and die and molding industries for coatings on semiconductor devices, food packaging, and cutting tools, etc., due to improvement in wear resistance, mechanical, chemical, optical properties, and electronic functionality by creating thin coatings with graded properties. The PVD process can be performed in different ways based on the method

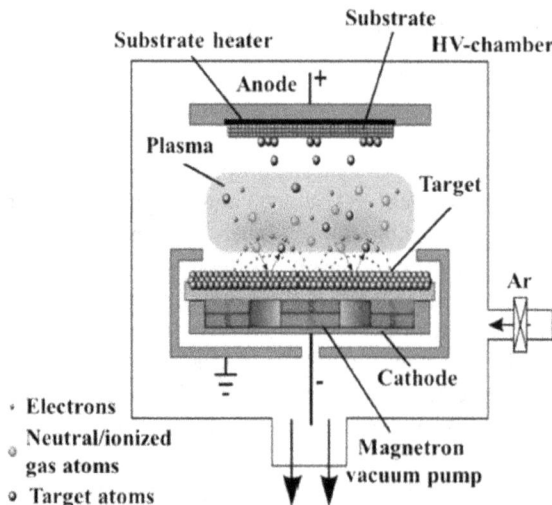

FIGURE 2.3 Schematic representation of the PVD process [2].

adopted to vaporize (atomize) the solid material, such as the sputtering-based PVD, evaporation-based PVD, and the plasma spray-based PVD, etc. [21]. All these PVD techniques require a complete vacuum for processing, and are performed in three to four steps, according to the type of material being used.

2.4.1.1 Evaporation-Based PVD

In this method the material desired to be deposited is evaporated by melting and vaporization by a suitable heat source. The heat can be supplied by hot filament, electrical resistance, electron beam, laser beam, or electric arc positioned at a distance from substrate material. Other heat source technologies are ion beam-evaporation, molecular beam, and a discharge supported evaporation process. The material in atomized form obtained by the vaporization of the target material is transferred to a substrate in a vacuum for coating. Thus the four fundamental steps for this technique are evaporation, transportation, reaction, and last deposition.

In the evaporation step, the target material is kept in a crucible followed by bombardment of high energy via a heat source, i.e. thermal energy by resistance heater, electron beam by electron-supplying system. This heating results in the increase in material temperature up to the melting point and vaporization. The vapor of the target material is transported to the substrate material and then a third or optional step, i.e. reaction, occurs. The reaction step is mandatory when the target material is a metal oxide, nitride, or carbide form. In this step reaction takes place between the atomized material and gases, such as methane, oxygen, or nitrogen during transportation before contacting the substrate. If the target or coating material is required in the pure form then the reaction step is avoided. Finally the last step is the deposition of the atomized coating or target material over the substrate.

In order to deposit functionally graded coating, two materials vaporized from different crucibles are simultaneously deposited over the substrate surface in a controlled manner based on the composition ratio of the desired material design. The main advantages of this PVD technique are the fast deposition rate, easy and simple use, conductor materials can be deposited on electronic circuits, and optical as well as dielectric coatings can be effortlessly attained. This PVD technique involves a number of disadvantages also, for example the evaporation coefficient of the coating material can be affected by surface contamination; the deposition rate can be affected by the geometry of the target-to-substrate material; the coating thickness is distance-dependent, thus it is not uniform; it is difficult to deposit target materials of higher melting points; it is difficult to achieve uniform heating, and there can be reactions of the heating container and target material.

2.4.1.2 Sputtering-Based PVD

Sputtering-based deposition techniques are available in various forms. These techniques can be categorized on the basis of the nature of the power sources utilized, which include diode or direct current sputtering, magnetron sputtering, radio frequency sputtering, and reactive sputtering. In sputtering-based techniques, ions of the sputtering gas (inert gas) with high kinetic energy are forced to continuously collide over the target material surface and result in the ejection of atoms. Subsequently these ejected atoms from the target metal are finally deposited over the substrate surface. The ions of the inert or sputtering gas are sped up and forced to hit the target or coating material surface. Then atoms from the target material surface are sputtered, or ejected, or removed, and then sputtered atoms flood from the chamber and are deposited over the substrate surface.

In this PVD technique four steps are involved: the sputtering of coating or target atoms to the gaseous phase; the transportation of sputtered atoms via plasma environment to the substrate surface; the reaction between the sputtered atoms and reactive gases (this step is voluntary); and finally atom deposition over the substrate surface. The transportation step includes the passing of target atoms via inert gas or sputtering, and plasma medium. Subsequently the electrons of the plasma medium collide with neutral atoms from inert gases, to create ions. The ejection rate of atoms driving out of the target material is employed to illustrate the yield of sputtering. This yield is usually in the range of 0.01–4 atoms/ion and depends on available energy of the sputtering gas ions, target material mass, incident angle of the sputtering gas ions, and the binding energy of the target material atoms.

The sputtering deposition method is very versatile and adaptable; it can deposit a variety of metals, alloys, compounds, and insulators. This process can perform in-situ cleaning of the substrate surface using a "back-sputtering" technique by a polarity reversal of the DC sputtering system, and influence the mechanical as well as electrical properties and adhesion quality of the coated material over the substrate. The PVD technique requires low-vacuum and lower line-of-sight deposition. The deposited material has a smaller grain with several orientations and less chance of material decomposition. This process offers effective control over material properties and lower shadow effect. Although this technique has several advantages, some drawbacks are also present, such as that a lower sputtering rate causes lower deposition, the majority of applied energies are transformed to heat, the emitted kinetic energy is low, the presence of gas inclusions in deposited material is common due to the low-vacuum process, and it is expensive to due to high capital investment.

2.4.1.3   Plasma Spray PVD

The plasma spray PVD technique was invented by Sulzer-Metco, and it is a hybrid deposition method. It combines low-pressure plasma spraying and a PVD technique, which allows the evaporation and deposition of the target or coating materials. The standard low-pressure plasma spraying technique is performed at 50–200 mbar pressure, and allows a coating deposition of 20 μm–1 mm thickness, wherein the plasma flame can stretch to 50–500 mm by reducing the pressure, allowing homogeneous and uniform coating deposition. In the plasma spray PVD, the pressure ranges from 0.5 to 2 mbar, and the plasma flame can become over 2 m long, with a diameter of 200–400 mm. The pressure in the plasma spray deposition is superior to the PVD process, but the high plasma temperature and plasma velocity (2000 m/s) allow feedstock to vaporize effortlessly, which permits easy coating development outside the stream [22]. Another technique wherein the material comes to the vapor phase while processing is chemical vapor deposition.

2.4.2 CHEMICAL VAPOR DEPOSITION TECHNIQUES

Chemical vapor deposition (CVD) is also a deposition method that is used to deposit functionally graded as well as thin-film coating. The CVD process is performed by keeping the coating material in a vacuum chamber, followed by vaporization due to decreasing the pressure of the circumference environment or heating until material evaporation. Meanwhile precursor gases are brought into the vacuum chamber having the heated substrate for coating, resulting in chemical reaction near as well as on the substrate surface. This reaction deposits a thin coating on the substrate and by-products are released out along with unutilized gases. The working of the CVD process is illustrated in Figure 2.4. The process occurs in cold and hot wall reactors, above atmospheric pressure, and at temperatures in the range of 200–1600°C. This technique is employed to create extremely pure materials exclusive of external defects by the deposition of desired material in a controlled way at the nanostructure or atomic level. With this technique, it is feasible to fabricate nanostructured, single-, or multilayered functionally graded coating having a precisely organized structure and dimensions. There are several types of improved CVD processes, such as the ions, the plasmas, the lasers, the hot filaments, the combustion reactions, or the photons. These processes are adopted with the CVD to boost deposition rates and/or to reduce deposition temperatures [23].

The CVD technique has several benefits, such as trouble-free coating in the undersides and the insides of geometrical aspects; high-aspect ratio holes can be filled entirely; applicability for a large variety of material with high purity; relatively superior deposition rates; and lesser requirement of vacuum than PVD processes. The CVD technique does not need a vacuum or high-grade electrical power supply and it results in low operating costs. Also there is no need to rotate the substrate to obtain wide-step coating; thus intricacies on certain coatings such as holes and slots can be easily achieved. The drawbacks of CVD include the following issues:

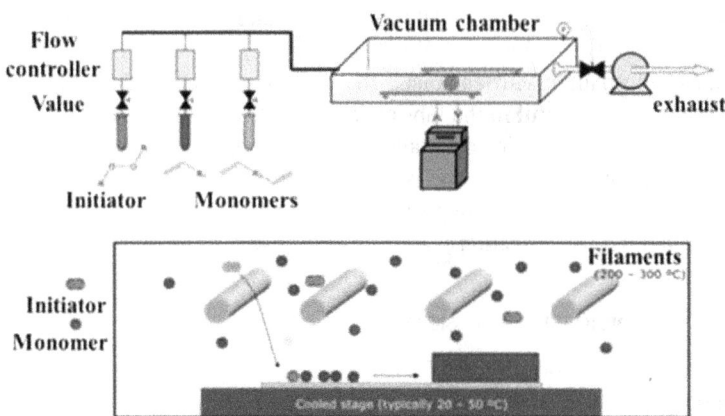

FIGURE 2.4 Schematic representation of the CVD process [2].

at room temperature precursors should be volatile and the precursors are explosive, corrosive, and toxic in nature. Additionally by-products from reactions of CVD are harmful, and several precursors are extremely costly. The deposition of films is generally performed at higher temperatures. The processing by CVD generates thermal stresses in coatings due to the difference of coefficients of thermal expansion. The preparation of coatings by CVD techniques with several metals is impossible [23].

2.5 DEPOSITION TECHNIQUES

The deposition techniques such as thermal spray and electrophoretic deposition are very useful for fabricating micro-scale thin FGM coatings. These methods are well capable to develop continuous or discrete gradient and multidimensional FGM coatings and are explained in the following subsections.

2.5.1 THERMAL SPRAYING

The thermal spray technique is an important process applied in the development of thin-dimension functionally graded surface coatings using a spraying process. The graded layer over the surface shields the parts because the surface needs proper safety against wear, corrosion, electrical, and thermal isolation. There are numerous methods adopted to fabricate FGMs with a gradient of material properties by the thermal spray coating process. The thermal spraying techniques using combustion phenomena are high-velocity oxy fuel spraying, flame spraying, and detonation guns. Also some techniques use electrical phenomena such as arc spray, vacuum plasma spray, atmospheric plasma spray, and low-pressure plasma spray. In this technique, initially a heat source is employed to melt the desired coating materials, followed by processing with gases and spraying over the substrate material. Subsequently materials in the molten state solidify and generate a solid layer as presented in Figure 2.5. Researchers have developed FGMs from ZrO_2/NiCoCrAlY, NiCrAlY/ yttria-stabilized zirconia (YSZ), TiO_2-hydroxyapatite (HA), and Ni/YSZ material systems using the thermal spray method. The developed materials showed uniform and defect-free coating with enhanced thermal as well as mechanical properties. This method is very useful in the fabrication of FGM coatings to enhance wear and corrosion resistance. Its main advantages include: lower heat input requirement, processing cost is low, production rate is high, and a wide variety of materials can be used for coating [24]. Another deposition technique, i.e. electrophoretic deposition, is detailed in the next subsection.

2.5.2 ELECTROPHORETIC DEPOSITION

The electrophoretic deposition (EPD) technique is used to make continuous as well as discrete gradient FGMs. It is a straightforward technique for fabricating properties gradient using the electrophoresis principle. This method is capable of developing difficult geometry at a low cost. First it was proposed in 1740 by G. M. Bose while performing a liquid siphon experiment. Mainly it contains two processes:

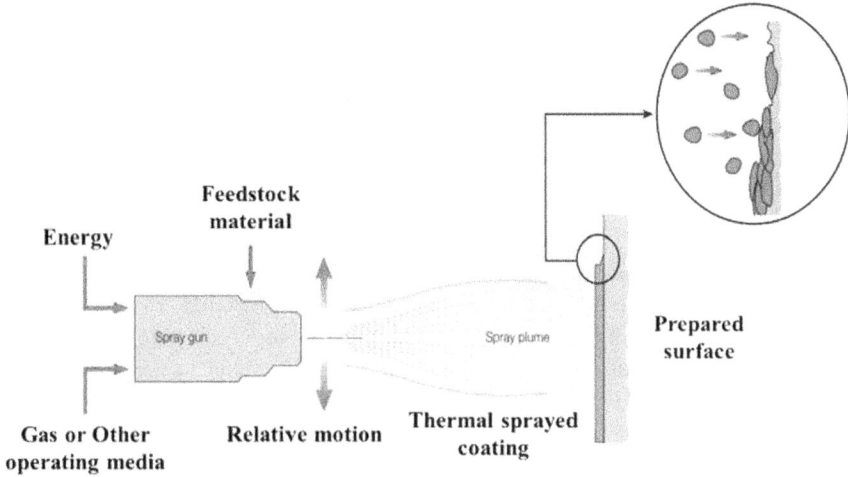

FIGURE 2.5 Representative diagram of thermal spraying [2].

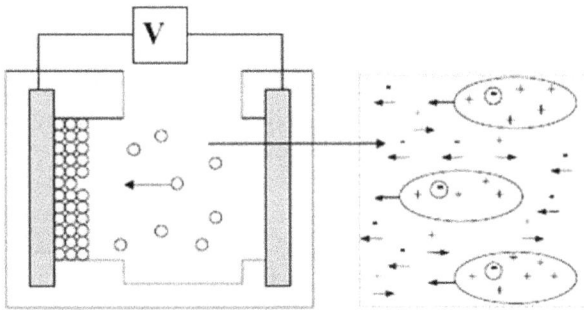

FIGURE 2.6 Representative diagram of the EPD technique [25].

electrophoresis and deposition as shown in Figure 2.6. Electrophoresis is a method to analyze the influence of electricity applied on suspension, which is the outcome of an applied electric field over motion of suspended particles. The accumulation of suspended particles is known as deposition. Thus to perform the EPD, it is essential that suspension particles should react to the employed electric field; also the particles must be stable – either electrosterically or electrostatically. Under the effect of chemical reaction or electrolysis, the active particles experience chemical reduction and deposit on the substrate surface to create a coating. Initially an interaction between the additives and solvent is employed to activate particles. Subsequently active particles go through the applied controlled electrical field. Finally, the particles develop an enlarged accumulation of the deposition on the electrode. The process equipment of EPD is easy to control and operate with accuracy; it is simple and low cost. The EPD process is suitable for numerous exercises, for example laminates, coatings (nm-mm), graded materials, textured, and infiltration material. This method proved very beneficial for preparing FGMs for Al_2O_3/YSZ, Al_2O_3/Ni, YSZ/Ni, and Al_2O_3/$MoSi_2$ material systems. This process is mainly used for coating purposes; its processing cost is lower and it is easy to manage the processing. Although it is useful for creating FGM coatings, the obtained gradient thickness is less and it is a time-consuming and environmentally harmful process [26].

2.6 SOLID PHASE PROCESSING TECHNIQUES

The solid-state production technique, i.e. powder metallurgy and friction stir processing, is a very effective method for the production of FGMs. Even though the powder metallurgy technique generates a discontinuous or discrete gradient, the high degree of control and the ability to fabricate bulk FGMs make this method suitable for numerous industrial applications.

2.6.1 POWDER METALLURGY TECHNIQUES

The powder metallurgy (PM) process is an old method of developing engineering product and nowadays due to its versatile capability has been adopted for the preparation of FGMs. It is a very prominent and feasible process for manufacturing bulk FGMs with discrete gradients. This method consists of four fundamental steps to generate FGMs: mixing, stacking, pressing, and sintering. The series of fundamental steps necessary for the preparation of gradients includes preparation and processing of the powder materials, stacking of powder mixture of different composition according to the design requirements, forming or compaction operations, and finally the consolidation or sintering to fulfill the service conditions requirement of the FGMs, as depicted in Figure 2.7. Powders of ceramic, metals, and alloys can be prepared in a range of crystallite sizes as well as distributions. The FGMs are fabricated using the powder particles of component materials. The powder particles are utilized have a composition or average particle size variation, according to the path stated by the design of the FGMs [27].

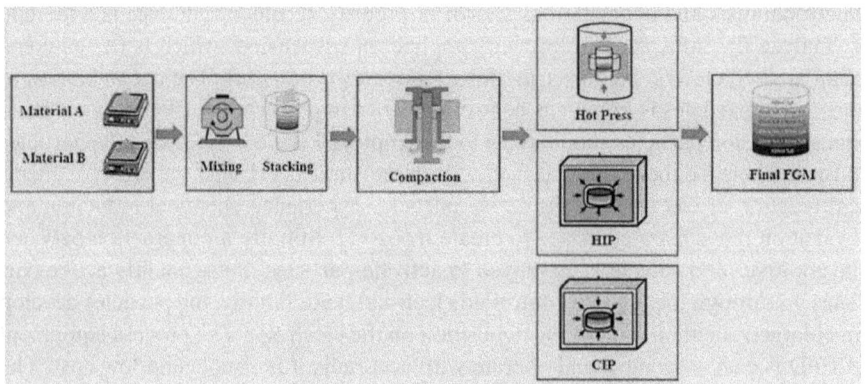

FIGURE 2.7   Representative diagram of powder metallurgy [2].

Therefore the first step in powder metallurgy is the mixing of powder particles of selected materials with designed volume fraction according to the desired composition gradient. The powder mixtures are stacked in the die according to a previously designed sequence to achieve a stepwise gradient. Subsequently the die with layered composition undergoes a compaction process and the output product is known as the green compact. The green compact is fragile in nature and requires further consolidation and goes through the sintering process. Sintering is a procedure to heat the green compact using a furnace in order to melt the elemental materials either partially or at the surface only, resulting in bonding between individual particles. In order to perform sintering, a high temperature is needed so the proper densification can be achieved by chemical reaction and diffusion process, wherein the thermodynamic factor plays a very crucial role. The challenging task while doing sintering is uniform consolidation of the green compact due to the presence of a material composition gradient. Non-uniform sintering results in frustum formation, warping, cracking, and splitting. Uniform sintering can be performed by incorporating an activating element along with the powder mixture to avoid particle size control. The activating element can be a transition metal such as copper, iron, etc. It is also possible to achieve uniform sintering by applying pressure simultaneously with sintering using the hot-pressing technique so a highly dense product can be obtained. Additionally sintering can be performed in three different ways, such as high-frequency induction heating, electric furnace sintering, and spark plasma sintering [28].

2.6.1.1 Spark Plasma Sintering (SPS)

SPS is a modern sintering technique utilizing plasma for the consolidation process. SPS involves direct heating and simultaneous pressing of stacked layers that causes internal heating, plasma generation, and a high sintering rate. The SPS process is a modification of the conventional sintering technique, wherein the powder compact is heated directly instead of separately in a furnace. In SPS, pulsed DC current passes from the die filled with the powder mixture, resulting in internal heating. Additionally pressure is also applied over the punch; thus by using pressure as well as temperature simultaneous high heating as well as sintering rate is achieved. As internal heating takes less time to achieve the desired sintering temperature and with small holding time, high sintering density is obtained. This method is very effective in the sintering of functionally graded green compact and the final product has high density. Additionally this method is very popular for the sintering of materials having high melting temperatures, such as cutting tool materials, biomaterials, etc., to achieve high fracture toughness, wear resistance, and bending strength. The biomaterial developed by SPS is utilized as bone implants and joint prostheses. Also the compliant pad used in the thermoelectric energy translation developed using Ni and Al via SPS has optimum composition gradient and densification [29].

In this method the powder mixtures are prepared according to the design requirement, and deposited in from of layer in the graphite die-punch assembly. Subsequently this die-punch arrangement is kept inside the SPS sintering chamber under controlled atmospheric conditions (argon/vacuum) between the movable punch. The punch moves according to the predefined compaction pressure and simultaneously

DC current is supplied to heat the complete assembly. Additionally cooling support is also provided for faster cooling according to requirements. Although this method offers various advantages, the size of the final product is limited due to the constraints of die size; also the set-up cost is very high, and more research is required for mass production [30].

Powder metallurgy is easy to operate, cheap, has near net-shape forming ability and better control over microstructure, and a composition gradient can be obtained. However, very intricate parts cannot be created using this process and the cost of the powder material is high. The key benefits of PM are the removal of cracks, macroscopic interface, and atomic defects, and the uniformity of composition with the absence of inter-metallic carbides and segregation. The precipitate phases formed due to processing are distributed uniformly. Friction stir processing is also a solid phase processing technique for FGM fabrication and is explained in the next subsection.

2.6.2 FRICTION STIR PROCESSING (FSP)

The operating principle of friction stir welding (FSW) is the foundation of the preparation of FGM by friction stir processing but the aim of each technique is different. FSW is the process of joining metals together, and in FSP the microstructure of the base material is modified by incorporating reinforcement. FSP involves a metallic tool with a threaded pin along with shoulder. This metallic pin is forced in the grooves over the base metal surface where reinforcement is required. The tool pin is able to rotate about its own axis; thus while processing the rotating tool pin is forced to the grove with the reinforcement material at a certain tilt angle. Subsequently the tool pin traverses along the designed path and frictional heat is created due to the movement, resulting in plastic deformation. The movement of the tool pin also results in the consolidation of material by hydrostatic pressure [31]. The first reported FGM fabrication was done on an AA5083/SiC material system, wherein SiC is reinforced in AA5083 in the composition gradient form. This method is mostly used in Al alloy, Mg alloy, zirconium alloy, and Cu alloy. Additionally it has proved extremely advantageous for nano or micro particle reinforcement, porosity gradient, and the alteration of casting microstructure [32].

The distinguishing advantages of FSP include accurate control, densification, homogeneity, uneven depth of processed zone, and microstructure refinement. Moreover, it requires less energy, is environmentally friendly, and does not alter the shape and size of the fabricated part. It moreover has the ability to perform the sintering of superior-quality products, but the size and quantities of fabricated sample are major constraints. Moreover, the processing as well as machinery cost is extremely high, and additional research effort is required to commercialize this process.

2.7 ADDITIVE MANUFACTURING PROCESSES

Additive manufacturing is also called rapid prototyping, rapid manufacturing 3D printing, direct-digital manufacturing, layered manufacturing, and solid-free form fabrication. In this method near net shape product can be directly manufactured without involving molds or the need for assembling or joining. Additionally additive manufacturing exhibits superiority over conventional manufacturing techniques in the preparation of geometrically complex product at low cost, with less time by permitting flexibility in design. The fast advancement of AM techniques means they are not any more limited in their application for materials having single-phase.

Nowadays, the AM process has become capable of processing materials with multiple phases and creates steady composition changes in structures, and this advanced processing technique of creating FGMs is named functionally graded additive manufacturing (FGAM). It symbolizes a layered preparation to progressively modify the composition of a component material to attain the required functionality, as shown in Figure 2.8. FGAM can fabricate three kinds of material variations: density variations in single-phase materials; composition variation in multi-phase materials; and deviations in the material composition as well as density together. The preparation of FGMs with composition or density gradient permits several properties. The variation of composition and/or density permits the fabrication of multifunctional FGMs with properties (such as graded thermal/magnetic properties/mechanical) that are presently unattainable through conventional manufacturing processes. In general, the FGAM process involves essential steps such as modeling (materials modeling, microstructural design, and geometrical modeling), slicing and simulation, manufacturing, and characterization.

There are several variants of AM technologies and according to the ASTM F42 board they are classified into seven categories. The seven categories are: material jetting (multi-jet modeling), vat photopolymerization (stereolithography), material extrusion (fused-deposition modeling), binder jetting (plaster-based 3D printing), powder-bed fusion (selective-laser sintering), sheet lamination (laminated objective manufacturing), and directed-energy deposition (laser-metal deposition). Mainly five essential steps are performed in any AM process. These steps are, first, the receipt of the CAD file in the machine; the second is the conversion of the CAD file into the additive manufacturing file (AMF) or standard triangulation language (STL); the

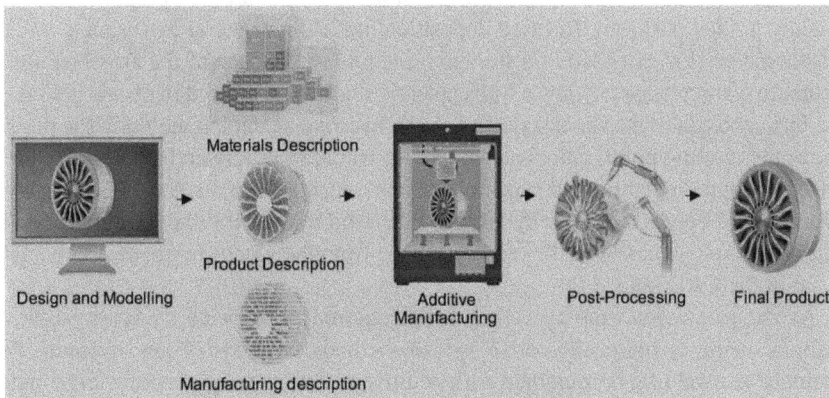

FIGURE 2.8   Representative diagram of additive manufacturing [33].

third is the slicing of the transformed file inside layers of a two-dimensional (2D) triangle, demonstrating 3D CAD data; and the fourth is layer-by-layer part preparation according to the path given in the 2D sliced CAD data; finally, there is the part removal from the AM setup for the finishing operation, such as heat treatment. The fourth step, i.e. building process, is very crucial and differentiates one AM process from another [34].

Several commercial software packages are available for the 3D printing of multi-material FGMs, such as Autodesk Monolith, voxel-based systems, and Grab CAD. These packages are not capable of fulfilling the industrial demand for desired graded properties due to the existing challenges such as inadequate understanding of multi-material systems, the absence of reliable technique, the inability to estimate final material properties, and higher costs. The FGM design is a crucial parameter for successful materials preparation, wherein the assignment and description of material characteristics and nature of voxel (a small unit of 3D volume) are the most significant parameters [33].

In material-extrusion AM, the building process is done by using a nozzle to deposit extruded material on the platform. According to the sliced data the first perimeter is deposited, then the area is filled by a selected raster pattern continuously layer-by-layer and finally fused together. Selective laser sintering AM is a laser-based method, used for materials in powdered form. It is also a layer-by-layer manufacturing process, wherein laser energy is focused on the powder layer and results in the fusion or sintering of particles together. The laser metal deposition process works on the direct energy deposition principle to create a discrete gradient, wherein the energy source and material are applied concurrently. This energy source heats the deposited material directly at the deposition site. Material is deposited in the form of wire or powder and energy is supplied in the form of a laser, e-beam, or arc [35]. Stereolithography works on photopolymerization and engages a light source to bond treatable resins with another material to develop a functionally graded component.

The material jetting technique is similar to a 2D inkjet printer. In this AM process the material is supplied on command into a structure platform. The desired materials are dropped on the base or ground and solidify layer by layer to prepare the functionally graded part [36]. In fused deposition modeling or metal extrusion a wax or plastic material is extruded in a storage place and the surface of the cross-sectional structure is traced layer by layer. Subsequently material is offered in plastic pellets or the form of a filament. The nozzle comprises resistive heaters to maintain the plastic above the melting point, and changes rapidly from the nozzle and creates the layer. Sheet lamination (laminated objective manufacturing) utilizes sheets coated with adhesive and bonded together by heated roller which sticks the sheets. Subsequently a laser is employed to slash the required shape from the glued sheets, as per the path given in the CAD data [37].

As the production time as well as costs required to fabricate FGMs by additive manufacturing is high, alternative hybrid methods have also been invented. For example, arc welding is combined with additive manufacturing to create large metal parts with graded structures at low production time and cost [38]. In addition plasma arc welding, plasma transferred arc, metal inert gas, and tungsten inert gas are also utilized. These types of techniques are described as wire and arc additive manufacturing because material is supplied as wire or powder and an arc is employed as power source.

Although there are numerous advantages of AM in fabricating FGMs, it is very difficult to get defect-free desired components due to following reasons: the large number of processing variables (such as powder flow rate, power density, substrate characteristics, displacement velocity, etc.) which have significant influence and are very difficult to optimize.

2.8 CHALLENGES AND FUTURE POTENTIAL IN FUNCTIONALLY GRADED MATERIALS FABRICATION

In the earlier sections, the manufacturing techniques of FGMs were elaborated in detail. Each technique displayed its credibility in generating a smooth gradient with enhanced material properties. This section details the gaps available in modern research on FGMs and directions for future work on the basis of the existing literature. The remarkable benefits of FGMs make them most noteworthy materials with significant future applications. Thus, researchers endeavor to develop state-of-the-art production techniques for the fabrication of FGMs to match the fourth industrial revolution. The fabrication techniques of FGM are differentiated by numerous advantages, such as the ability to manufacture complex shapes with graded material properties, reduction in thermal stresses, and control over deformation and wear, as well as corrosion rate. Although FGMs offer several exciting material properties, their application has numerous challenges, including quality control of gradient, mass production with accuracy and precision, and cost.

In the PVD technique it is necessary to prepare the particles of optimum size and analyze the residual stresses to obtain the required quality gradient. In the CVD process also residual stresses result in detrimental effects on graded material properties; thus effective numerical analysis is crucial. In the EPD process still extensive study on the gradient formation mechanism and characterization is required. The numerical simulation to analyze the influence of residual stress on the material gradient in the thermal spray process needs to be explored. A database for optimum process parameters of the powder metallurgy process for a variety of materials is still not available. In the centrifugal casting process detailed study on the choice of optimal mass fraction, particle size and shape, bonding between particles and matrix, and particle movement is lacking. Slip as well as tape casting requires effort to develop variety of gradient with newer material systems and analysis on the bonding among particles and matrix. Infiltration casting necessitates research on the analysis of bonding between particles and matrix, optimal particle size, and the influence of residual stress. The state-of-the-art additive manufacturing process is a potential fabrication technique with the ability to decrease economic burden in the production system with flexibility in the required shape as well as size. In spite of immense abilities, still a number of issues must be solved. It is recommended to improve the performance of AM processes by detailed material characterization, comprehensive database, and feedback control for varieties of FGMs.

Abundant research work has been carried out by distinguished researchers since the origin of the FGM. The available research work is dedicated to the design and modeling of FGMs. Also extensive studies focused on the fabrication as well as characterization of the FGMs also exist. This outstanding sophisticated material has enormous prospects to modernize the manufacturing industries in the twenty-first century. But still the main setback exists in terms of the manufacturing cost of FGMs due to the high cost of powder development as well as processing methods. The FGM gradient preparation is a crucial step of FGM development; and it can be improved using numerical modeling by forecasting the material combination as well as properties which facilitate reduction in the overall cost of fabrication. Upgradation of laboratory research to an industrial level for appropriate technology transfer is also recommended due to the critical need for cost reduction and the fabrication of FGMs. Also reproducibility in gradation, properties, and geometry of FGMs is an important research criterion for effective production techniques. It is trusted that by fulfilling the proposed research objectives, the production cost of FGMs will go down, and the reliability in the gradient and fabrication process will improve.

2.9   CONCLUSIONS

The concept of FGMs offers enormous potential to bring a revolution in the field of materials science and manufacturing by having diverse application areas such as defense, engineering, biomedical transport, and petroleum; but its development is the main challenge. Although plenty of methods are available, its commercialization is very limited due to issues related to quality control and reproducibility. The fabrication of FGMs involves knowledge as well as skills of different areas, e.g. engineering, chemistry, and physics. Thus to convert the idea of gradient into reality along with accurateness as well as precision, two essential aspects are extremely crucial. The first essential aspect is quantitative aspect assessment of composition, microstructure, and properties gradient to accomplish desirable features. The other important aspect is selection fabrication techniques for the preparation of required gradients with accuracy and precision using the data available from the quantitative analysis. The approach that must be adopted for the preparation of application-oriented composition gradients is known as materialization, which involves material design followed by fabrication and characterization. In order to successful fabricate FGMs, the following important aspects should be considered:

(i)   The selection of the most appropriate processing method according to the area of application, shape of the final product, and available material combination as well as properties data.

(ii)   The application of numerical analysis tools to achieve optimum processing parameters and complete control over the final material properties.

(iii)   Performance evaluation of the prepared component using the appropriate method and further improvement using obtained results.

Although several processing methods are available for FGM development, still there is a requirement for advancement over the current state for mass production in terms of proper feedback control and reliability along with repeatability and accurateness, cost-efficient and superior value, and fault-less parts. In this chapter existing processing methods are discussed in brief with their field of application, ability, and limitations. Additionally several research gaps along with related future potentials are also mentioned, which will be helpful for researchers to explore the concept of gradient in effective ways.

REFERENCES

1. Li W, Han B. Research and application of functionally gradient materials. *IOP Conf Ser Mater Sci Eng* 2018;394:022065. https://doi.org/10.1088/1757-899X/394/2/022065.
2. Saleh B, Jiang J, Fathi R, Al-hababi T, Xu Q, Wang L, et al. 30 years of functionally graded materials: An overview of manufacturing methods, applications and future challenges. *Compos B Eng* 2020;201:108376. https://doi.org/10.1016/j.compositesb.2020.108376.
3. Parihar RS, Setti SG, Sahu RK. Recent advances in the manufacturing processes of functionally graded materials : A review. *Sci Eng Compos Mater* 2016;25:309–36. https://doi.org/10.1515/secm-2015-0395.
4. Jamian S, Watanabe Y, Sato H. Formation of compositional gradient in Al/SiC FGMs fabricated under huge centrifugal forces using solid-particle and mixed-powder methods. *Ceram Int* 2019;45:9444–53. https://doi.org/10.1016/j.ceramint.2018.08.315.
5. Rajan TPD, Jayakumar E, Pai BC. Developments in solidification processing of functionally graded aluminium alloys and composites by centrifugal casting technique. *Trans Indian Inst Met* 2012;65:531–7. https://doi.org/10.1007/s12666-012-0191-0.
6. Rahimipour MR, Sobhani M. Evaluation of centrifugal casting process parameters for in situ fabricated functionally gradient Fe-TiC composite. *Metall Mater Trans B* 2013;44:1120–3. https://doi.org/10.1007/s11663-013-9903-z.
7. Vajd A, Samadi A. Optimization of centrifugal casting parameters to produce the functionally graded Al–15wt%Mg2Si composites with higher tensile properties. *Int J Meteorol* 2020;14:937–48. https://doi.org/10.1007/s40962-019-00394-1.
8. El-Galy IM, Saleh BI, Ahmed MH. Functionally graded materials classifications and development trends from industrial point of view. *SN Appl Sci* 2019;1:1–23. https://doi.org/10.1007/s42452-019-1413-4.
9. Inaguma Y, Sato H, Watanabe Y. Fabrication of Al-based FGM containing TiO2 nanoparticles by a centrifugal mixed-powder method. *Mater Sci Forum* 2010;631–632:441–7. https://doi.org/10.4028/www.scientific.net/MSF.631-632.441.
10. Watanabe Y, Inaguma Y, Sato H, Miura-Fujiwara E. A novel fabrication method for functionally graded materials under centrifugal force: The centrifugal mixed-powder method. *Materials* 2009;2:2510–25. https://doi.org/10.3390/ma2042510.
11. Watanabe Y, Shibuya M, Sato H, Miura-Fujiwara E, Kawamori S. Fabrication of Mg-based functionally graded materials by a reaction centrifugal mixed-powder method. *Keikinzoku/J Jpn Inst Light Met* 2012;62:153–9. https://doi.org/10.2464/jilm.62.153.
12. Kunimine T, Sato H, Miura-Fujiwara E, Watanabe Y. New processing routes for functionally graded materials and structures through combinations of powder metallurgy and casting. *Adv Funct Graded Mater Struct* 2016,33–48. https://doi.org/10.5772/62393.

13. Kunimine T, Shibuya M, Sato H, Watanabe Y. Fabrication of copper/diamond functionally graded materials for grinding wheels by centrifugal sintered-casting. *J Mater Process Technol* 2015;217:294–301. https://doi.org/10.1016/j.jmatprotec.2014.11.020.

14. Szymański, OE, Tokarski T, Kurtyka P, Drożyński D, Żymankowska-Kumon S. Reactive casting coatings for obtaining in situ composite layers based on Fe alloys. *Surf Coatings Technol* 2018;350:346–58. https://doi.org/10.1016/j.surfcoat.2018.06.085.

15. Watanabe Y, Inaguma Y, Sato H. Cold model for process of a Ni-aluminide/steel clad pipe by a reactive centrifugal casting method. *Mater Lett* 2011;65:467–70. https://doi.org/10.1016/j.matlet.2010.10.042.

16. Jabbari M, Bulatova R, Tok AIY, Bahl CRH, Mitsoulis E, Hattel JH. Ceramic tape casting: A review of current methods and trends with emphasis on rheological behaviour and flow analysis. *Mater Sci Eng B* 2016;212:39–61. https://doi.org/10.1016/j.mseb.2016.07.011.

17. Luo G, Dai Y, Liu S, Yu K, Shen Q, Zhang L. Fabrication and properties of W-Cu functionally graded material by tape-casting. *Key Eng Mater* 2014;616:66–71. https://doi.org/10.4028/www.scientific.net/KEM.616.66.

18. Chmielewski M, Pietrzak K. Metal-ceramic functionally graded materials: Manufacturing, characterization, application. *Bull Pol Acad Sci Tech Sci* 2016;64:151–60. https://doi.org/10.1515/bpasts-2016-0017.

19. Hsu HC, Chou JY, Tuan WH. Preparation of AlN/Cu composites through a reactive infiltration process. *J Asian Ceram Soc* 2016;4:201–4. https://doi.org/10.1016/j.jascer.2016.03.003.

20. Rahvard MM, Tamizifar M, Boutorabi MA., Shiri SG. Effect of superheat and solidified layer on achieving good metallic bond between A390/A356 alloys fabricated by cast-decant-cast process. *Trans Nonferrous Met Soc China* 2014;24:665–72. https://doi.org/10.1016/S1003-6326(14)63109-5.

21. Rajak DK, Wagh PH, Menezes PL, Chaudhary A, Kumar R. Critical overview of coatings technology for metal matrix composites. *J Bio- Tribo-Corros* 2020;6:12–30. https://doi.org/10.1007/s40735-019-0305-x.

22. Shahidi S, Moazzenchi B, Ghoranneviss M. A review-application of physical vapor deposition (PVD) and related methods in the textile industry. *EPJ Appl Phys* 2015;71:1–13. https://doi.org/10.1051/epjap/2015140439.

23. Choy KL. Chemical vapour deposition of coatings. *Prog Mater Sci* 2003;48:57–170.

24. Espallargas N. *Introduction to Thermal Spray Coatings*. Elsevier Ltd. 2015. https://doi.org/10.1016/B978-0-85709-769-9.00001-4.

25. Put S, Vleugels J, Anné G, Van der Biest O. Functionally graded ceramic and ceramic–metal composites shaped by electrophoretic deposition. *Colloids Surf A Physicochem Eng Asp* 2003;222:223–32. https://doi.org/10.1016/S0927-7757(03)00227-9.

26. Lee SH, Woo SP, Kakati N, Kim DJ, Yoon YS. A comprehensive review of nanomaterials developed using electrophoresis process for high-efficiency energy conversion and storage systems. *Energies* 2018;11:3122–3203. https://doi.org/10.3390/en11113122.

27. Kawasaki A, Watanabe R. Concept and P/M fabrication of functionally gradient materials. *Ceram Int* 1997;23:73–83. https://doi.org/10.1016/0272-8842(95)00143-3.

28. Parihar RS, Gangi Setti S, Sahu RK. Effect of sintering parameters on microstructure and mechanical properties of self-lubricating functionally graded cemented tungsten carbide. *J Manuf Processes* 2019;45:498–508. https://doi.org/10.1016/j.jmapro.2019.07.025.

29. Strojny-Nędza A, Pietrzak K, Węglewski W. The influence of Al2O3 powder morphology on the properties of Cu-Al2O3 composites designed for functionally graded materials (FGM). *J Mater Eng Perform* 2016;25:3173–84. https://doi.org/10.1007/s11665-016-2204-3.

30. Mitsuo N, Narushima T, Nakai M. *Advances in Metallic Biomaterials: Tissues, Materials and Biological Reactions.* Springer. 2015. https://doi.org/10.1007/978-3-662-46842-5.
31. Deepak D, Sidhu RS, Gupta VK. Preparation of 5083 Al-SiC surface composite by friction stir processing and its mechanical. *Int J Mech Eng* 2013;3:1–11.
32. Hangai Y, Suto S, Utsunomiya T. Fabrication of aluminum sandwich panel with functionally graded foam consisting of dissimilar aluminum alloys. *Keikinzoku/J Jpn Inst Light Met* 2017;67:595–6. https://doi.org/10.2464/jilm.67.595.
33. Loh GH, Pei E, Harrison D, Monzón MD. An overview of functionally graded additive manufacturing. *Addit Manuf* 2018;23:34–44. https://doi.org/10.1016/j.addma.2018.06.023.
34. Yan L, Chen Y, Liou F. Additive manufacturing of functionally graded metallic materials using laser metal deposition. *Addit Manuf* 2020;31:100901. https://doi.org/10.1016/j.addma.2019.100901.
35. Li X, Tan YH, Willy HJ, Wang P, Lu W, Cagirici M, et al. Heterogeneously tempered martensitic high strength steel by selective laser melting and its micro-lattice: Processing, microstructure, superior performance and mechanisms. *Mater Des* 2019;178:107881. https://doi.org/10.1016/j.matdes.2019.107881.
36. Yap YL, Wang C, Sing SL, Dikshit V, Yeong WY, Wei J. Material jetting additive manufacturing: An experimental study using designed metrological benchmarks. *Precis Eng* 2017;50:275–85. https://doi.org/10.1016/j.precisioneng.2017.05.015.
37. Singh S, Ramakrishna S, Singh R. Material issues in additive manufacturing: A review. *J Manuf Process* 2017;25:185–200. https://doi.org/10.1016/j.jmapro.2016.11.006.
38. Rodrigues TA, Duarte V, Miranda RM, Santos TG, Oliveira JP. Current status and perspectives on wire and arc additive manufacturing (WAAM). *Materials* 2019;12:1121–1162. https://doi.org/10.3390/ma12071121.

3 Liquid Phase Processing Techniques for Functionally Graded Materials

Lalit Ranakoti, Brijesh Gangil, and Shashikant Verma

CONTENTS

3.1 INTRODUCTION

This chapter sheds light on the processing of graded materials without discussing the utility and significance of functionally graded materials (FGMs). In the mid-1990s, the evolution of FGM processing techniques was initiated, and since then, the development in the methods of manufacturing FGMs has received much recognition. The processing of FGMs can be carried out via several techniques, namely: (a) liquid state processing, (b) solid-state processing, and (c) deposition processing [1–3]. The

DOI: 10.1201/9781003097976-3

aforesaid manufacturing techniques have sub-categories according to the variation in material property requirements, such as the base metal state, process temperature, pressure, applied centrifugal forces, etc. Also, which method should be adopted for the processing of FGMs depends on their final use in real-life situations [4–7]. These manufacturing techniques have their respective benefits and drawbacks with particular significance for the analysis point of view, but liquid state manufacturing is gaining much recognition for its ability to produce good-quality FGMs.

3.2   LIQUID STATE PROCESSING OF FGMS

Manufacturing by liquid state processing consists of various techniques such as slip casting, tape casting, infiltration, and centrifugal casting, processed at relatively low cost, and it can yield continuous structure with graded properties [8]. However, the process parameters are generally difficult to control, resulting in wettability and complications in processing molten metals.

3.2.1   CENTRIFUGAL CASTING

Centrifugal casting is a method of fabricating FGMs in bulk mainly influenced by the driving force of centrifugal action that arises because of the rotation given to the mold during the time of fabrication [9]. The mold's rotation causes the material to flow in an outward radial direction, resulting in the formation of two or more than two phases across the thickness. In this technique, casting is carried out in a liquid state only for the manufacturing of dedicated materials for specific purposes [10]. The first patent in centrifugal casting of graded composite was filed by A. G. Echardt in the year 1809. Later, in 1858, the initiation of commercialization of centrifugal casting took place in which polymer pipes were manufactured [11–13]. In the early 1990s, the centrifugal casting method attained full proficiency by producing graded composites in bulk.

The difference in the density of constituents of graded composites such as filler, fiber, and matrix causes the formation of different phases at the core and periphery [14]. This method is initiated by mixing particulates/fillers and base metal in a separate beaker followed by the pouring of the mixture in the circular mold subjected to centrifugal forces caused by the rotation, as shown in Figure 3.1. After finishing with the rotation of the mold, the circular mold is left to cool. The casted product is taken out after solidification, which generally takes 16 to 24 hours. The graded composite obtained via centrifugal casting is predominantly affected by various parameters such as the rotating speed of the mold, filler type, filler size (length/diameter size), filler weight percentage, melting temperature, cooling rate, and coating [15]. Centrifugal casting can be carried out for the fabrication of small parts, called vertical centrifugal casting, and for long parts, called inline horizontal centrifugal casting [16]. Depending on the matrix-fiber rapport, centrifugal casting is further divided into two types: (a) the centrifugal solid particle method (CSPM) and (b) the centrifugal in-situ method (CISM).

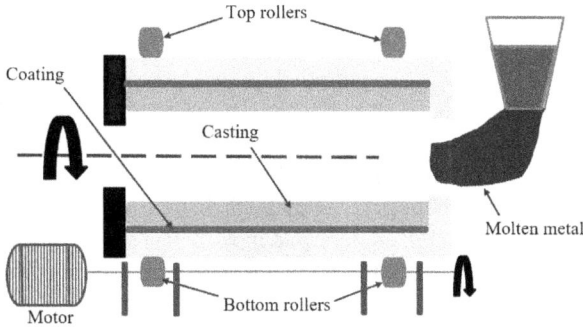

FIGURE 3.1   Centrifugal casting of FGMs.

3.2.1.1   Centrifugal Solid Particle System (CSPM)

Generally, tube-type structures with gradually increasing density of constituents are produced by CSPM. Reinforcement with a melting point higher than the matrix material is used in this method where reinforced particles remain in solid form during the processing, and the distribution takes place along the radius, depending upon the density of the reinforcement [17]. Matrix materials used in this technique are magnesium, aluminum, copper, and their alloys, while Al_2O_3 and SiC are the chief reinforcing material used in this technique [18]. The combination of these matrices and reinforcement has enormous scope in engineering applications. As per the recent finding reported in the literature, the mold rotational speed was between 600–2500 rpm to optimize process parameters and to achieve optimum mechanical and tribological properties. It is to be noted that the properties of graded composite were enhanced up to 1500 rpm, but as the speed passed 1500 rpm the properties of the graded composite started to decline due to the clustering of filler particles leading to deviation of fabrication from the desired objective [19]. However, a precise rotational speed of 1200 rpm was claimed in the recent research for obtaining satisfactory graded composite with adequate distribution of particles [20]. The parameters for deciding the weight percentage of the filler materials mainly depend on their physical properties and chemistry with the base material. The range of particle weightage is generally taken as 2.5 to 20% of the matrix material [21]. This technique is very flexible as an adjustment in parameters can be made to get the desired properties.

3.2.1.2   Centrifugal in-Situ Method (CISM)

In this type of manufacturing of FGMs, the reinforced particulates generally have a melting point higher than the base material, which is used to fabricate the tubular section. Centrifugal force is applied on the mold while it is in the cooling stage, i.e. solidification, and not during the fabrication process, leading to the formation of a graded structure before reaching the state of crystallization [22]. FGM materials manufactured with this method majorly depend upon the densities of reinforcement particles and matrix, but the particles' shape and size also significantly affect the properties of FGMs. The weight fraction of reinforcement can be increased to 40 wt.% in this method. This method produces graded composites with higher tensile strength as obtained by CSPM. This technique can be used at higher gravitational force as the chances of the splitting of reinforcement reduce due to the mold's rotation at the time of solidification [23]. Preheating the mold is beneficial in this method as the reinforcement in the base metal is distributed evenly compared to the distribution of reinforcement in an unheated mold [24]. The reinforcement materials used in this technique are alloys of aluminum, copper, magnesium, and silicon.

3.2.2 CENTRIFUGAL SLURRY POURING METHOD

Particles with different densities are reinforced in this technique. This method uses the concept of powder metallurgy and centrifugal casting. The particles are placed inside the rotating mold already containing liquid in the suspended form [25]. The rotation of the mold causes the formation of a graded surface due to the difference in mass of constituents. Upon completion of the process, the liquid slurry is separated from the graded composite. Fabricated graded composite then undergoes sintering for finishing [26]. A significant drawback of this method, as observed during the graded composite processing, is the accumulation of reinforcement having high mass at the mold's bottom. Thus, the graded characteristics are impossible to achieve across the whole thickness of the composite. It was reported that approximately 35% of the graded composite thickness was formed in a graded structure by this method [27]. Later, this issue was resolved by introducing slurry 100 mm thick in the rotating mold which helps in the uniform dispersion of particles across the whole thickness even without sedimentation as shown in Figure 3.2a.

3.2.3 CENTRIFUGAL PRESSURIZED METHOD

In this method, the desired graded structure is developed before the application of centrifugal effect on the mold containing the reinforcement and matrix [28]. A simple pressure is applied on the mold by the centrifugal force to achieve the required

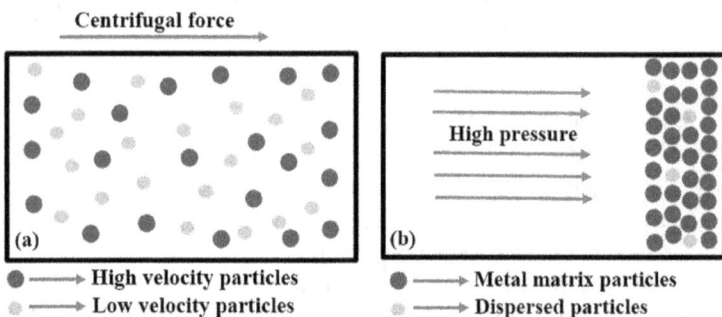

FIGURE 3.2 (a) Centrifugal slurry casting and (b) centrifugal pressurized casting.

graded composite as shown in Figure 3.2b. It is subdivided in three categories viz: (a) the mixed centrifugal power method, (b) the centrifugal sintered casting method, and (c) the reactive centrifugal casting method.

3.2.3.1 Mixed Centrifugal Power Method (MCPM)

Graded composites with a small particle size are difficult to fabricate as small particles face high resistance against the molten metal movement. Since small particles have low momentum due to low mass and as reported in Stoke's law theory, reducing particle size results in a decrease in velocity [29]. The low momentum of particles affects their distribution and spreads them in a more uneven format. In the MCPM method, centrifugal force is given to the mold after a particular grading is attained at nano scale. This method asserts that nano-sized particles can effortlessly reinforce the base metal and the desired grading structure [10]. Composites of titanium oxide, copper oxide, and magnesium silicates can be easily reinforced with various base metals.

3.2.3.2 Centrifugal Sintered Casting Method (CSCM)

As discussed in MCPM, the gradation in the structure is obtained before the start of the centrifugal process. It is not favorable for some kinds of powders; thus a new technique called CSCM emerged. This process combines both centrifugal sintering and centrifugal casting. Initially, a ring-type structure of metal and matrix dispersed with particles is made [30]. The predeveloped amalgam of metal and particles is inoculated in the rotating mold provided that the melting point of the dispersed particles is higher than that of the base metal for the manufacturing of FGMs. In the meantime, sintering is executed for melting by the action of centrifugal forces. Molten metal is then added under the action of centrifugal force to obtain the required FGM. The melting and heating of metal matrix particles lead to the formation of graded surfaces in the form of a ring-type structure [31]. A typical schematic diagram of CSCM is shown in Figure 3.3.

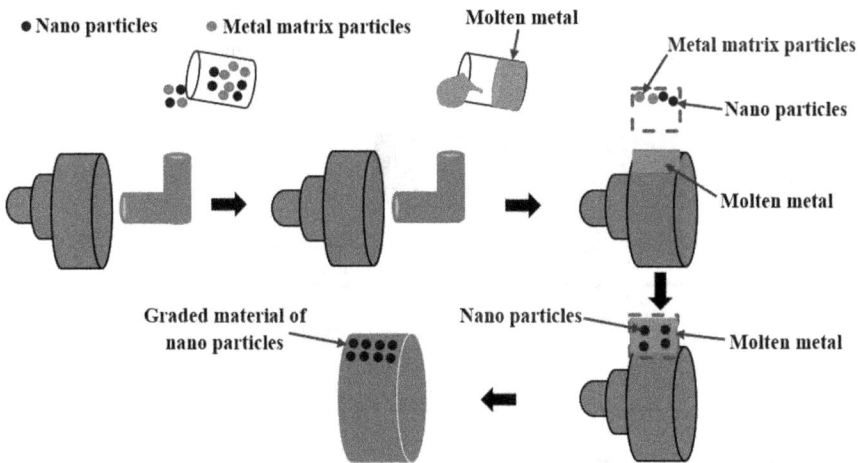

FIGURE 3.3   Schematic diagram of centrifugal sintering casting method.

3.2.3.3   Reactive Centrifugal Casting Method (RCCM)

In reactive casting, an exothermic reaction takes place between the constituents of the graded composite, which generates an intermetallic compound of a very high melting point. Thus external heating is not required in this process [32]. By this method, smooth and rust-free graded surfaces of iron, nickel, and aluminum can be easily manufactured, which have applications in the storage containers of the oil and gas industries.

3.2.4   Slip Casting Method

Capillary action plays a major role in this technique for the fabrication of graded surfaces as fine grains of ceramic particles fall on the permeable mold by capillary action. The passage of draining is a slippery surface of clay, and that's why the process is called slip casting [33]. After achieving a certain thickness, the left-over material is drained out through passages. Changing the grain size from fine to coarse will result in a change of the chemical composition of layers of graded structure. The product obtained by the slip casting method has different porosities at different layers of the graded structure. At high temperatures, the slip casting method fails as the base metal starts dispersing at a very high rate. This process has the ability to fabricate materials of complex shapes of a continuous graded system [34]. The fabrication of cast tape by slip casting is shown in Figure 3.4.

3.2.5   Tape Casting Method

This method is used in the production of ultra-thin FGMs with a gradient layer ranging from 50 to 100 micrometers. This work was initially used for FGMs of conventional ceramics material but, later on, was extended for complex ceramics. This method consists of the casting of layers of substrate by dispersing ceramic powder in a mixture of binder, solvent, and plasticizers. This method is generally employed to carry out large-scale production. Integrated circuits of electron accessories and multi-layer condensers are fabricated by this method [35]. One such type of process

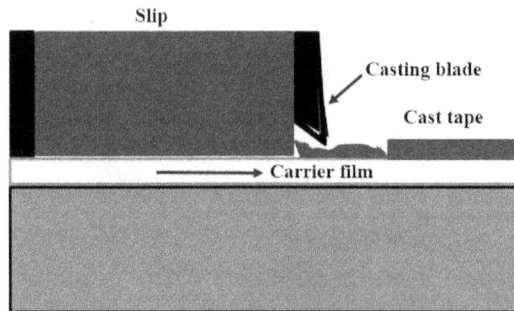

FIGURE 3.4   Fabrication of cast tape by slip casting.

is known as doctor blading in which a ceramic in the form of powder is poured on a moving layer supported on a flat surface usually made of polymer binder. The slurry thus formed is stretched by a sharp knife of desired thickness. The solvent evaporates, and the layer formed falls on a disk under processing. Products with enhanced thermal stability and comparatively low interfacial defects are cast by this method [36].

3.2.6 Method of Infiltration

This method is liquid state casing in which particles are soaked in the vacant space of a molten matrix. The particles are infiltrated to occupy the dispersed phase of molten metal. The structure thus formed is known as functionally graded material in dispersed form [37]. The infiltration can be achieved either by capillary action or by applying external pressure as shown in Figure 3.5. The interface of molten metal undergoes chemical changes resulting in the formation of FGMs. This is a time-saving method and has impressive accuracy in production [38]. The product produced by this method possesses good electrical characteristics.

3.2.7 Gel Casting

Gel casting is used to create complex shapes of functionally graded ceramic materials of porous and dense structures. A colloidal solution is formed by the polymerization of monomers caused by the mixing of free radical initiators. The forming time is short in this process, and it is capable of quickly producing a large number of products. The apparatus required to carry out gel casting is of low cost which makes it economical too. The process is followed by mixing powder of ceramics in a solution containing cross-linker, catalyst, monomer, and initiator. The mixing leads to the formation of a slurry, which is then poured into a pre-defined mold where in-situ polymerization occurs and results in forming a gel of polymer and water. The gel is removed from the mold in wet form and left to cool. The binder is removed from the gel after it dries, and it is finally subjected to sintering. The various processes followed in gel casting for the fabrication of FGMs are shown in Figure 3.6.

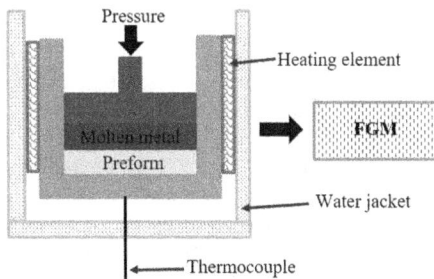

FIGURE 3.5 Fabrication of FGM by pressurized infiltration.

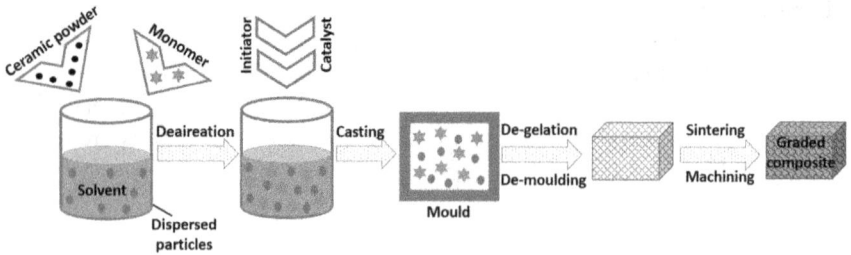

FIGURE 3.6   Process involved in gel casting of FGM.

3.3   CONCLUSIONS

The manufacturing of functionally graded materials can be accomplished swiftly by liquid phase processing. The centrifugal method is assumed to be the most common but conventional method by which several materials can be casted easily. Particles at micro scale and nano scale can be easily reinforced for producing FGMs by the centrifugal casting method. Ultra-thin graded materials can be manufactured by slip casting while powdery materials which are difficult to fabricate by slip casting can be processed by tape casting. To achieve a graded material in dispersed form, the infiltration technique is employed. These manufacturing methods aim to provide the manufacturer with guidance to help them choose the FGM process that is most consistent with their technological standards and answer their economic-related questions.

REFERENCES

1. Kieback, B., Neubrand, A., Riedel, H. Processing techniques for functionally graded materials. *Materials Science and Engineering A*, 2003. 362(1–2): 81–106.
2. Naebe M., Shirvanimoghaddam K. Functionally graded materials: A review of fabrication and properties. *Applied Materials Today*, 2016. 5: 223–45.
3. Gangil, B., Patnaik, A., Kumar, A. Mechanical and wear behavior of vinyl ester-carbon/cement by-pass dust particulate filled homogeneous and their functionally graded composites. *Science and Engineering of Composite Materials*, 2013. 20(2): 105–116.
4. Bohidar, S. K., Sharma, R., Mishra, P. R. Functionally graded materials: A critical review. *International Journal of Research*, 2014. 1(4): 289–301.
5. Gangil, B., Patnaik, A., Kumar, A., Kumar, M. Investigations on mechanical and sliding wear behaviour of short fibre-reinforced vinylester-based homogenous and their functionally graded composites. *Proceedings of the Institution of Mechanical Engineers Part L: Journal of Materials: Design and Applications*, 2012. 226(4): 300–315.
6. Gangil, B., Patnaik, A., Kumar, A., Biswas, S. Thermo-mechanical and sliding wear behaviour of vinyl ester–cement by-pass dust particulate-filled homogenous and their functionally graded composites. *Proceedings of the Institution of Mechanical Engineers Part J: Journal of Engineering Tribology*, 2013. 227(3): 246–258.
7. Ranakoti, L., Rakesh, P. K. Physio-mechanical characterization of tasar silk waste/jute fiber hybrid composite. *Composites Communications*, 2020. 22:100526.
8. Gasik, M. M. Functionally graded materials: Bulk processing techniques. *International Journal of Materials and Product Technology*, 2010. 39(1–2): 20–29.

9. Rathee, S., Maheshwari, S., Noor Siddiquee, A., Srivastava, M. Distribution of reinforcement particles in surface composite fabrication via friction stir processing: Suitable strategy. *Materials and Manufacturing Processes*, 2018. 33(3): 262–269.

10. Saleh, B., Jiang, J., Ma, A., Song, D., Yang, D. Effect of main parameters on the mechanical and wear behaviour of functionally graded materials by centrifugal casting: A review. *Metals and Materials International*, 2019. 25(6): 1395–1409.

11. Watanabe, Y., Inaguma, Y., Sato, H., Miura-Fujiwara, E. A novel fabrication method for functionally graded materials under centrifugal force: The centrifugal mixed-powder method. *Materials*, 2009. 2(4): 2510–2525.

12. Saleh, B., Jiang, J., Fathi, R., Al-hababi, T., Xu, Q., Wang, L., Song, D., Ma, A. 30 years of functionally graded materials: An overview of manufacturing methods, applications and future challenges. *Composites Part B: Engineering*. 2020:108376.

13. Chumanov, I. V., Anikeev, A. N., Chumanov, V. I. Fabrication of functionally graded materials by introducing wolframium carbide dispersed particles during centrifugal casting and examination of FGM's structure. *Procedia Engineering*, 2015. 129: 816–820.

14. Pradeep, A. D., Rameshkumar, T. Review on centrifugal casting of functionally graded materials. *Materials Today: Proceedings*. 2020.

15. El-Galy, I. M., Ahmed, M. H., Bassiouny, B. I. Characterization of functionally graded Al-SiCp metal matrix composites manufactured by centrifugal casting. *Alexandria Engineering Journal*, 2017. 56(4): 371–381.

16. Lin, X., Liu, C., Xiao, H. Fabrication of Al–Si–Mg functionally graded materials tube reinforced with in situ Si/Mg2Si particles by centrifugal casting. *Composites Part B: Engineering*, 2013. 45(1): 8–21.

17. El-Hadad, S., Sato, H., Watanabe, Y. Wear of Al/Al3Zr functionally graded materials fabricated by centrifugal solid-particle method. *Journal of Materials Processing Technology*, 2010. 210(15): 2245–2251.

18. Jamian, S., Watanabe, Y., Sato, H. Formation of compositional gradient in Al/SiC FGMs fabricated under huge centrifugal forces using solid-particle and mixed-powder methods. *Ceramics International*, 2019. 45(7): 9444–9453.

19. Savaş, Ö. The production and properties of Al3Ti reinforced functionally graded aluminum matrix composites produced by the centrifugal casting method. *Materials Research Express*, 2019. 6(12): 126532.

20. Owoputi, A. O., Inambao, F. L., Ebhota, W. S. A review of functionally graded materials: Fabrication processes and applications. *International Journal of Applied Engineering Research*, 2018. 13(23): 16141–16151.

21. Sharma, N. K., Bhandari, M., Dean, A. Applications of functionally graded materials (FGMs). *International Journal of Engineering Research & Technology*, 2014. 2(3).

22. Watanabe, Y., Kurahashi, M., Kim, I. S., Miyazaki, S., Kumai, S., Sato, A., Tanaka, S. I. Fabrication of fiber-reinforced functionally graded materials by a centrifugal in situ method from Al–Cu–Fe ternary alloy. *Composites Part A: Applied Science and Manufacturing*, 2006. 37(12): 2186–2193.

23. Watanabe, Y., Kim, I. S., Fukui, Y. Microstructures of functionally graded materials fabricated by centrifugal solid-particle andin-situ methods. *Metals and Materials International*, 2005. 11(5): 391–399.

24. Oike, S., Watanabe, Y. Development of in-situ Al-Al2Cu functionally graded materials by a centrifugal method. *International Journal of Materials and Product Technology*, 2001. 6(1–3): 40–49.

25. Jayachandran, M., Tsukamoto, H., Sato, H. Watanabe, Y. Formation behavior of continuous graded composition in Ti-ZrO2 functionally graded materials fabricated by mixed-powder pouring method. *Journal of Nanomaterials*, 2013. 2013: 1–8.

26. Kinoshita, K., Sato, H., Watanabe, Y. Development of compositional gradient simulation for centrifugal slurry-pouring methods. *Materials Science Forum*. Trans Tech Publications Ltd, 2010. 631: 455–460.

27. Sobczak, J. J., Drenchev, L. Metallic functionally graded materials: A specific class of advanced composites. *Journal of Materials Science & Technology*, 2013. 29(4): 297–316.

28. Edwin, A., Anand, V., Prasanna, K. Sustainable development through functionally graded materials: An overview. *Rasayan Journal of Chemistry*, 2017. 10: 149–152.

29. El-Wazery, M. S., El-Desouky, A. R. A review on functionally graded ceramic-metal materials. *Journal of Materials and Environmental Science*, 2015. 6(5): 1369–1376.

30. Kunimine, T., Shibuya, M., Sato, H., Watanabe, Y. Fabrication of copper/diamond functionally graded materials for grinding wheels by centrifugal sintered-casting. *Journal of Materials Processing Technology*, 2015. 217: 294–301.

31. Ebhota, W. S., Karun, A. S., Inambao, F. L. Principles and baseline knowledge of functionally graded aluminium matrix materials (FGAMMs): Fabrication techniques and applications. *International Journal of Engineering Research in Africa*. Trans Tech Publ Ltd, 2016. 26: 47–67.

32. Jayakumar, E., Jacob, J. C., Rajan, T. P. D., Joseph, M. A., Pai, B. C. Processing and characterization of functionally graded aluminum (A319)—SiC p metallic composites by centrifugal casting technique. *Metallurgical and Materials Transactions A*, 2016. 47(8): 4306–4315.

33. Peng, X., Yan, M., Shi, W. A new approach for the preparation of functionally graded materials via slip casting in a gradient magnetic field. *Scripta Materialia*, 2007. 56(10): 907–909.

34. Lopez-Esteban, S., Bartolomé, J. F., Pecharroman, C., Moya, J. S. Zirconia/stainless-steel continuous functionally graded material. *Journal of the European Ceramic Society*, 2002. 22(16): 2799–2804.

35. Zhong, Z., Zhang, B., Jin, Y., Zhang, H., Wang, Y., Ye, J., Liu, Q., Hou, Z., Zhang, Z. Ye, F. Design and anti-penetration performance of TiB/Ti system functionally graded material armor fabricated by SPS combined with tape casting. *Ceramics International*, 2020. 46(18): 28244–28249.

36. Gitis, V., Rothenberg, G. *Ceramic Membranes: New Opportunities and Practical Applications*. Wiley. 2016.

37. Jedamzik, R., Neubrand, A., Rödel, J. Functionally graded materials by electrochemical processing and infiltration: Application to tungsten/copper composites. *Journal of Materials Science*, 2000. 35(2): 477–486.

38. Pompe, W., Worch, H., Epple, M., Friess, W., Gelinsky, M., Greil, P., Hempel, U., Scharnweber, D. Schulte, K. J. M. S. Functionally graded materials for biomedical applications. *Materials Science and Engineering A*, 2003. 362(1–2): 40–60.

4 Gaseous Phase Processing Techniques for Functionally Graded Materials

*Praveennath G. Koppad, M.R. Ramesh,
S. Joladarashi, S.T. Aruna, Nagaraja C. Reddy,
and C. Siddaraju*

CONTENTS

DOI: 10.1201/9781003097976-4

4.1 INTRODUCTION

The need for advanced engineering materials to work in harsh operating conditions for challenging applications like wear-resistant linings, rocket heat shields, plasma facings for fusion reactors, durable construction systems, biocompatible devices, and bone tissue reproduction is ever increasing. Engineers have been trying to address such concerns by seeking answers from nature. Nature has produced many beautiful materials like plants, trees, skin, bone, and teeth. There are many other materials which seem to be simple superficially but when studied thoroughly are found to have complex structures to serve multiple functions. One beautiful example that can be seen in nature is bamboo which belongs to the grass family known as Bambuseae. The hollow aerial stems of bamboo which grow to a diameter of 20–25 cm and height of 30 m have the capability to bear their own weight and bending loads resulting from flowing winds. The multi-scale structure (i.e. hierarchical) of bamboo is shown in Figure 4.1 and consists of strong crystalline cellulose in lignin matrix in the form of closed-cell foam and bundles of fibrils [1]. One can view bamboo as a fiber reinforced composite material in which the properties are varying in the radial direction. The brown spots of fiber bundles seem to be closely packed in at the outside radial direction while in the interior they are widely dispersed. As mentioned the fibrils are densely dispersed only at the periphery of the bamboo and their primary function is to provide stiffness. Due to the density, size, and number of fibril bundles and the structure of culms the mechanical and physical properties of bamboo vary considerably, characterizing them in the class of functionally graded materials.

If we consider the human body it has numerous systems which contain several sub-systems which are meant to carry out multiple functions. Take the example of the skin of the human body which contains a complex multilayered system, namely the hypodermis, the dermis, and the epidermis. Each layer has distinct properties that are required to perform different functions. The inner-layer hypodermis provides the

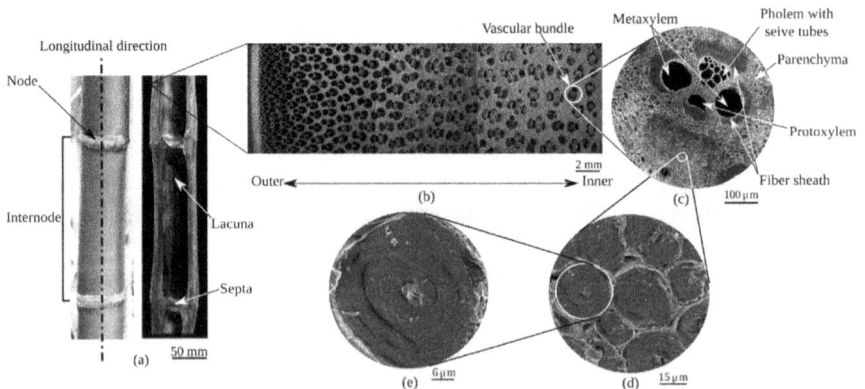

FIGURE 4.1 Internal structure of bamboo: (a) cross-section, (b) graded structure, (c) closely packed fiber bundles consisting of fibrils, (d) fibrils having lameller structure, and (e) cellulose-hemicellulose-lignin building blocks [1].

main structural support to the skin and aids in shock absorption. The middle layer known as the dermis consists of glands which prevent the skin from drying out and regulate body temperature. The outer layer of the skin, the epidermis, is made up of scale-like cells, creating our skin tone and protecting against water loss. It is interesting to note that the thickness of the epidermis depends on its location on the body. For example on the eyelids the thickness is half a millimeter while on the palms and soles it is more than 1 millimeter. It is important to observe how nature has made the skin which macroscopically is a single system but its multiple layers perform different functions. Another interesting example is the tendon-to-bone insertion site; why is this interesting? Let's see. Here, the first thing to note is that bone is a stiff material while tendon is a soft compliant material. So strictly from a stiffness point of view there is a huge mismatch because the modulus of the tendon (transverse and longitudinal moduli), which is in the range of 45 to 450 MPa, while that of bone is 20 GPa. Now one can expect a high level of stresses at the interface between these two materials. However, both these materials are separated by, or to say at the interface they contain, unique tissue known as enthesis whose function is to transfer stress from the soft tendon to stiff bone. Despite failure-inducing stress concentrations the enthesis shows exceptional durability and bears loads which are equivalent to multiple orders of magnitude of weight of the body. This is mainly because of its distinct fiber organization and composition which is just below 1 mm or in some cases it is ~500 μm [2]. Further at the tendon-to-bone insertion site, the stress is transferred effectively due to two gradation mechanisms [3]. First, the varying mineral content from bone to tendon material, and second, variation in the orientation of collagen fiber. Because of such change in the composition of tissue from tendon to bone, there is an efficient transfer of load. Along with the load transfer, the graded structure or in simple words continuous change in tissue composition right from the tendon-to-bone insertion site allows complex multi-axis bending. In this section each case of natural system or structure, whether it be bamboo or skin or tendon-to-bone insertion site, is provided in detail to understand how these materials contain several sub-systems with individual functionality. From an engineering point of view this as an interesting scenario for scientists and engineers as they can derive solutions for various engineering problems. As mentioned earlier, the skin or tendon-to-bone insertion site might get damaged and in such circumstances the job of scientists and engineers is to repair the system or to recreate a similar system which can mimic the original one. In such cases it is very important to have the knowledge of these naturally occurring functionally graded materials.

4.1.1 BRIEF BACKGROUND

It is seen that many applications need a material that properties have to be different in a different direction. In such materials the composition and structure tend to vary in one direction or over the entire volume. This new type of material was termed functionally graded materials (FGM), which are quite similar to composite materials but have continuous or discontinuous variations in the compositions and microstructures. The term gradient describes non-uniformity in the morphology and size of the materials and represents the transition between the constituents of the materials. The concept of FGM is not new and was proposed earlier by Japanese researchers in the early 1980s. The development of new material was targeted towards reduction in thermal stresses and high-temperature performance for rocket engines by providing a grading interface rather than a sharp interface which caused failure in traditionally used composite laminates. However, the official terming as functionally gradient materials was done in the year 1986 and a long-term research program was initiated in Japan for hypersonic spacecraft application. The project required a material which was capable of sustaining a high temperature of 2000 K and another requirement was to inhibit the transmission of such high temperatures to the other parts of the material. After 1987, the concept and results obtained through various FGM research programs were disseminated through international symposiums and international exchange programs. The first symposium on FGM was conducted in the year 1988 by a dedicated forum on FGMs. The objectives of this forum were to publicize the research findings through conferences and workshops, survey the possible applications, and assist in technology transfer from research institutes to the industry. In the year 1990 the first international conference on FGM was held in Sendai, Japan (FGM 1990) which was followed by regular conferences held every two years [4]. In the year 2001 an international workshop presenting the recent trend and forecast was conducted under the chairmanship of Prof. Naotake Ooyama. Various topics like modeling and simulation, automatic manufacturing systems for FGM, residual stress measurement, ultrasonic imaging, and the biocompatibility of FG implant materials were presented. Since then regular research programs, international symposiums, and workshops have been held across the world. The latest update to this is that the 16th international conference on FGM (FGM 2021) will be conducted in the year 2021 in Hartford, USA.

4.1.2 ORGANIZATION OF CHAPTER

The chapter starts with an introduction section which describes how scientists and engineers draw inspiration from nature to initiate work on FGM. A brief background on how the research work started on FGM, different initiatives are undertaken by government and private agencies, and the dissemination of research work through conferences and workshops are presented. The current status of research work on FGM in mechanical engineering fields is presented in Section 4.2. In Section 4.3, the processing techniques via the gaseous phase route are provided along with broad classification. The major classes like thermal spray, and physical and chemical vapor deposition are discussed along with research work covered in these areas. Important deposition techniques like atmospheric plasma spray, suspension plasma spray, high-velocity oxy-fuel spraying, vacuum plasma spray, electron beam physical vapor deposition, pulsed laser deposition, plasma-enhanced/assisted CVD, and metal-organic CVD are covered. Section 4.4 covers the computational modeling and analysis part in which the prediction of thermophysical properties and thermal stress distribution using numerical and simulation approaches is presented. Section 4.5 presents the applications of functionally graded coating (FGC) in the thermal barrier and biomedical fields. The chapter concludes by summarizing the benefits of FGC over conventional coatings for a wide variety of applications.

4.2   CURRENT STATUS OF RESEARCH

Many development works are being carried out on the effective processing of FGM with desired gradient or layers to meet end applications. It is well known that high-temperature and biomedical applications like the gas turbine, mining, and medical industries are in constant need of novel materials with enhanced performance and expected to meet design life. A tremendous amount of work is being carried out to develop FGM coatings to replace coatings developed by conventional techniques. As of now many new combinations are being attempted and investigated for use in bio-medical applications such as bone implants and thermal barrier coatings for gas tur-bine applications. The deposition of coatings is carried out using the latest techniques like vacuum plasma spraying, flame spraying, cold spraying, atmospheric plasma spraying, and solution precursor plasma spraying. For example functionally graded Al/Al_2O_3 coatings developed using high-pressure cold spraying are preferred for space applications due to their low solar radiation absorption and comparatively high infrared emittance. The combination of a metallic bond coat, CoNiCrAlY/NiCrAlY, and a ceramic coat, partially stabilized zirconia, with a graded structure has been tested to withstand high-temperature erosion and corrosion in the gas turbines of aero-engines [5, 6]. If the example of biomedical applications is considered then there is a dire need for material which is not only wear-resistant and biocompatible but is also high strength, durable, and corrosion resistant. In such cases biologic coat-ings comprised of functionally graded structures are more favorable. Such coatings are developed using novel thermal spray techniques like high-velocity suspension flame spraying (HVSFS) and suspension plasma spraying (SPS). These techniques are capable of depositing very fine particles in the sub-micrometer or nanometer range. The integration of different materials with a wide range of size results in graded structures which possess more merits then conventional biologic coatings. Such novel coatings would enhance durability, reduce fatigue phenomena in ser-vice conditions, prohibit the release of metal ions from the metal substrate, enhance osteointegration, and minimize the possibility of infection [7]. Another potential area where FGM is considered as a potential candidate material is in the transforma-tion of conventional impervious pavements into pervious pavements. The objective behind this type of application is to have an environmentally friendly transportation medium and the implementation of water management systems. In this regard open graded friction course (OGFC) mixtures have attracted many potential investors due to their stone-to-stone contact and good connectivity air void content. The stone-to-stone contact imparts resistance to disintegration and plastic deformation while the air void content helps in drain-ability and noise reduction. Although OGFC graded structures have a number of benefits, there are still certain drawbacks which limit their wide application [8]. Even the concrete industry is drawing some inspiration from naturally occurring FGM and putting an effort into minimizing energy con-sumption, the optimal use of cement, and improving the lifespan by developing new types of materials like layered concrete and continuously graded concrete. In the case of layered concrete beams the benefits are attained via weight and cement consumption reduction. On the other hand continuously graded concrete is still under analytical research, and studies have shown weight reduction ranging from 27% to 62% for the cases of beams and structural walls. Along with weight reduction, better thermal performance is achieved by choosing varying concrete compositions. In order to fabricate graded concrete and its structures, various techniques like fresh-on-fresh casting, spray deposition, and additive manufacturing are employed [9, 10]. Currently, FGMs are tried out in advanced applications where conventional materials are used to address performance and design life-related issues.

4.3 PROCESSING TECHNIQUES

FGMs can be in the form of bulk material or thin coatings whose main objective is to replace sharp interfaces with a gradually changing interface to survive harsh working environments. One can play with the composition and microstructure to design an FGM suitable for the requirements of an application. There are many methods to develop FGMs but the scope of the present chapter is limited to only gaseous phase processing techniques. The major deposition techniques like thermal spray, physical vapor deposition, and chemical vapor deposition will be the primary focus of this chapter. These deposition techniques are further classified into different methods to meet the specific requirements of applications and are shown in Figure 4.2. From now onwards rather than FGM the new abbreviation used will be FGC, that is functionally graded coatings. When the concept of coatings comes to mind then it is quite

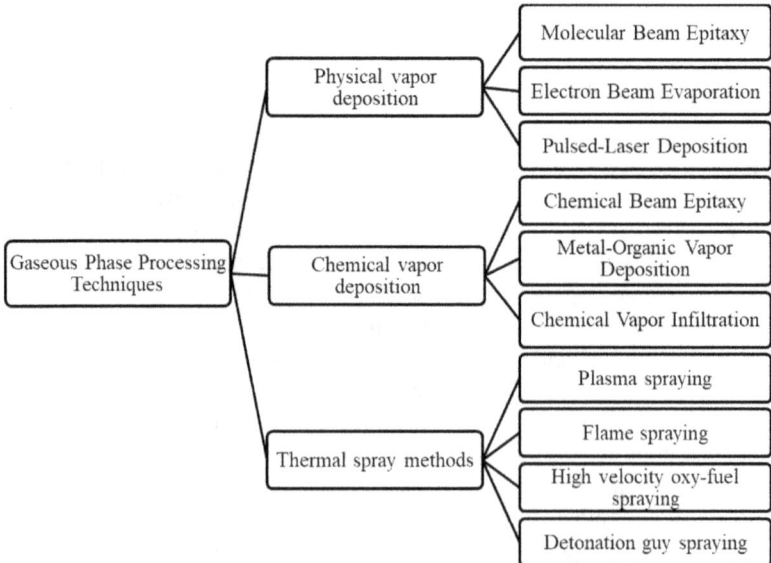

FIGURE 4.2   General classification of gaseous phase processing techniques.

clear that the objective is to protect a base material from wear, corrosion, oxidation, or erosion. Why use FGC when conventional coating materials are available? The answer to this question is that conventional coatings can perform only one function but FGCs are designed in such a way that they can perform multiple functions with prolonged life.

4.3.1 THERMAL SPRAY DEPOSITION

Thermal spray is the process in which premixed powder combinations in the form of powder, wire, or rod are introduced to a plasma flame or other heat source like laser beams and arcs. For plasma spray the feedstock material is generally in the form of powder, and these powder particles undergo partial or complete melting in the plasma flame. The powder in semi-molten or molten state is deposited onto the surface with the help of gas pressure. After impinging the powder particles undergo plastic deformation and rapid solidification, thus forming a deposit via successive impingement. A simple schematic depicting an example of thermal spraying of FGM on an incinerator burner nozzle is shown in Figure 4.3 [11]. This method is used to produce protective coatings of polymers, metals, and ceramics on different types of substrate materials. The benefits of thermal spray techniques include versatility in terms of materials that can be sprayed, coatings of varied thickness sometimes up to 10 mm that can be produced, high deposition rates, low environmental impact, and low processing costs. Thermal spray is classified into three major processes, plasma arc, electric arc, and flame, which further contain several subsets. Important subsets of thermal spraying processes are arc spraying, high-velocity oxygen fuel (HVOF) spraying, flame spraying, plasma spraying, suspension plasma spraying (SPS), atmospheric plasma spraying (APS), and special class, low-pressure cold spraying

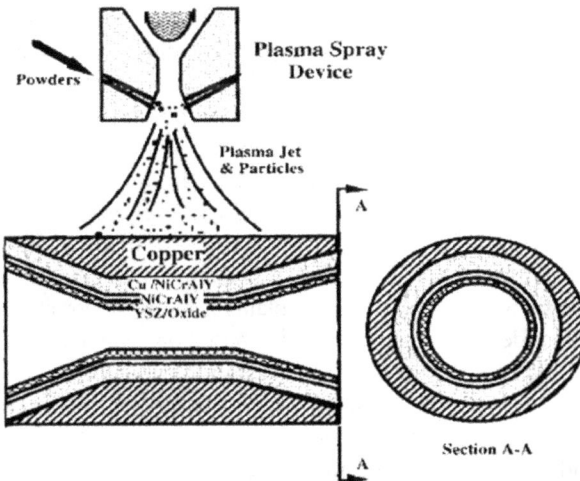

FIGURE 4.3   Schematic of thermal spraying of FGC on burner nozzle application [11].

(LPCS), and high-pressure cold spraying (HPCS). The first immediate application of FGM explored was thermal barrier coatings and the latest application is biologic coatings for orthopedic implants. The need for FGC on substrate over conventional thermal spray coatings was due to the fact that in the latter high mechanical and thermal stresses developed at the interface resulting in poor performance and premature failure. The design and development of coatings with gradient composition and structure are an attempt to address such concerns [12, 13].

Table 4.1 provides different FGC combinations developed for various applications using different thermal spray techniques in the most recent years [14–23]. Right from reduction in thermal stresses in the thermal barrier coatings for hot components of gas turbines to promoting osteointegration and minimizing infection by choosing biologic coatings for orthopedic implants are covered in Table 4.1. It is important to note that these are only a few examples of FGC and are provided here to understand the importance of thermal spray which is employed for a wide variety of applications.

4.3.1.1   Atmospheric Plasma Spraying (APS)

This technique is one of the most versatile thermal spray techniques as it can be applied for a wide variety of coating materials on various substrates including metallic ones. Everything from pure metals, ceramics, and alloys to advanced materials like cermets, carbides, superalloys, self-fluxing alloys, MCrAlY materials, and abradable materials can be sprayed using this technique. One can achieve various surface functionalities like electrically conductive surfaces, thermally conductive surfaces, anti-skid surfaces, high friction surfaces, lubricious surfaces, restoration of worn surfaces, fretting-resistant surfaces, cavitation-resistant surfaces, and alkaline- and saline-resistant surfaces using the APS process. As of now APS is used in pump components, gas turbine components, steam turbine components, propellers, shafts, mold release coatings, household irons, thread guides, anilox rolls, corona rolls, and impression rolls.

Xia et al. [24] reported the development of hydroxyapatite coatings on medical-grade Ti-6Al-4V alloy meant for orthopedic applications. An atmospheric plasma spraying (make: F4-MB, Sulzer Metco, Switzerland) system with optimal spray parameters like spraying distance of 100 mm and powder feeding rate of 25 rpm was chosen for the deposition of coatings. Developed coatings were subjected to a degradability test in Tris-HCl buffer solution for 14 days and osteogenic differentiation analysis. Fabricated coatings showed a high level of crystallinity, enhanced stability, and reduction in degradation rate. The formation of a nanorod structured surface of hydroxyapatite coating as shown in Figure 4.4 helped in improving the osseointegration between the bone and implants [24]. Rezapoor et al. [25] reported the functionally graded wear-resistant coating based on Fe and TiC on CK45 carbon steel substrate. The CK4 steel is generally used in the manufacturing of ship shafts, wheel hubs, and transmission parts which usually suffer from wear-related problems. The graded coating of 105 μm thickness had a top layer of TiC while in-between layers were made of Fe–TiC composites using an atmospheric plasma spray system (make: METCO-TB3). The adhesion and wear tests were conducted as per ASTM C633 and ASTM G99 standards respectively. For the wear test the pin-on-disc test

TABLE 4.1

List of Authors Published on FGMs Developed Using Thermal Spray Technique in the Last Five Years

Sl. No.	FGM	Author and Year	Thermal Spray Technique	Applications
1	AISI 410 steel/tungsten	Matejicek et al. [14], 2021	Argon-shrouded plasma spraying	Plasma-facing components used in fusion reactors
2	Hydroxyapatite/Al_2O_3	Singh et al. [15], 2020	Atmospheric plasma spray	Load-bearing clinical applications and to promote early bone tissue growth
3	$La_2(Zr_{0.7}Ce_{0.3})_2O_7$/YSZ	Taleghani et al. [16], 2020	Atmospheric plasma spray	Thermal barrier coatings on the hot sections of gas turbines
4	SiC/ZrB_2	Torabi et al. [17], 2020	Argon-shrouded plasma spraying	Hypersonic flight vehicles, re-entry vehicles and nozzles
5	NiCoCrAlYTa	Mendoza et al. [18], 2020	Combustion flame spray	Thermal barrier coating and hot corrosion resistance applications
6	Tungsten/ EUROFER	Emmerich et al. [19], 2020	Vacuum plasma spraying	Protection of first-wall (FW) structures of fusion power plants
7	Mullite/ZrO_2 and mullite/La_2O_3	Shreeram et al. [20], 2018	Transferred arc plasma spray	Thermal barrier coatings for internal combustion engine
8	Hydroxyapatite/titania	Henao et al. [21], 2018	High-velocity oxygen fuel spray	Bioactive coatings to improve the osseointegration of metallic implants
9	NiCrAlY/ZrO_2	Vakilifard et al. [22], 2017	Atmospheric plasma spraying	Hot section parts of gas turbines, like blades and vanes
10	YSZ/$Gd_2Zr_2O_7$	Carpio et al. [23], 2017	Atmospheric plasma spraying	Thermal barrier coatings to protect the metallic parts of gas turbines

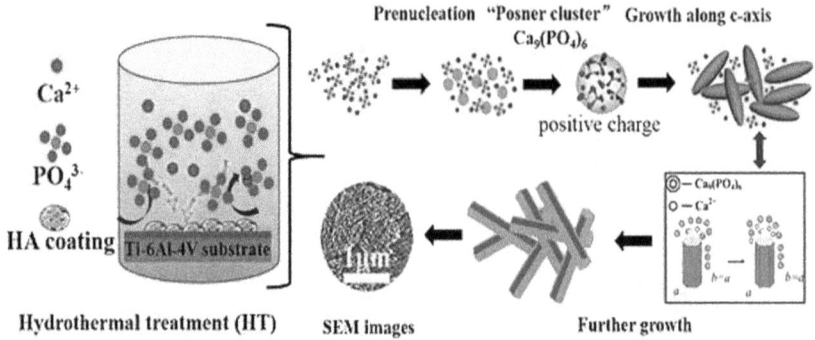

FIGURE 4.4 Mechanism explaining the formation of nanorod structured hydroxyapatite coatings [24].

rig with a counterbody made up of high carbon chromium alloy steel of 66 HRC hardness was employed. Higher adhesion strength and lower weight loss compared to other coatings were observed for functionally graded Fe–TiC coatings. Different systems like Al@NiCr/8YSZ [26] and nano-YSZ–alumina [27] for thermal barrier applications and nano-structured hydroxyapatite [28] and TiO_2-hydroxyapatite [29] for biomedical applications developed using APS have been reported.

4.3.1.2   High-Velocity Oxy-Fuel (HVOF)

In this spraying technique the oxygen and fuel are mixed together and introduced in the combustion chamber where they are ignited. The gas formed in the chamber is at very high pressure and temperature and when ejected through a nozzle it tends to travel at supersonic speeds. The feedstock powder is fed into this gas stream and propelled at the substrate at extremely high velocity, hence the name HVOF. Due to the higher particle velocity, dense and hard coatings with minimal porosity can be obtained. High adhesion strength, thicker coatings due to lower residual stresses, higher hardness due to lower degradation of carbide phases, and retention of powder chemistry due to high-velocity propelling of powder particles are some of the main advantages of the HVOF technique. A wide variety of material including stainless steel, chromium carbide, tungsten carbide, and many other carbides, oxides, and nitrides can be sprayed using this technique.

The HVOF technique is not only used for metallic and ceramic coatings but also used to deposit polymeric composite coatings for erosion barrier applications [30]. High-temperature polyamide and WC–Co were chosen as the matrix and reinforcing phase for the development of composite coating. The functionally graded coatings were obtained by the HVOF spraying of polymer composite material followed by the wire arc spraying of zinc and finally a top coating of WC–Co was provided. The FGC architecture developed using a combination of HVOF and wire-arc techniques is shown in Figure 4.5. The HVOF parameters chosen were a spray distance of 150 mm, powder feed rate of 3 g/min, surface speed of 0.11 m/s, substrate temperature of 230°C, and nitrogen carrier gas flow rate of 0.5×10^{-4} m^3/s. The

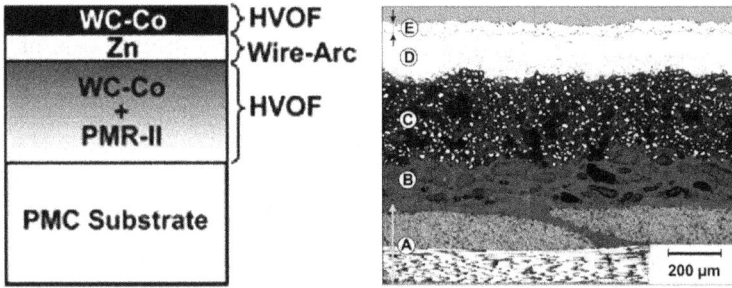

FIGURE 4.5 Architecture of FGC showing (a) polymer composite substrate, (b) polyamide layer, (c) polyamide/WC–Co layer, (d) Zn layer for binding, and (e) WC–Co top coat layer [30].

effectiveness of coatings as an erosion barrier was tested through statistical analysis of erosion test results using design of experiments. Erosion tests conducted using air jet erosion tester (Koehler Instrument Company, Inc., Bohemia, NY, USA) showed that FGCs were insensitive to erodent whose angle of incidence was in the range of 20° to 90° and temperature increase from 20 to 250°C. This erosion-resistant property was mainly due to the transitional metallic layer of zinc provided in the coating which not only provided sufficient protection to the underlying structure but was effective in preventing the polymer composite from being exposed to erodent. The HVOF technique was also used for the development of a thermal barrier coating and in the improvement of tribological performance for different applications like traction drives, cams, gears, and tappets used in transportation systems. Stewart et al. [31] reported the development of functionally graded WC–NiCrBSi coatings for such applications. Rolling contact fatigue tests conducted to study the combined effect of rolling and sliding motion showed that the FGC vacuum-treated at 1200°C showed improved performance due to the formation of complex carbides. Taking one step further, Henao et al. [20] employed the HVOF technique to produce functionally graded hydroxyapatite/titania coatings for biomedical applications. An in-vitro test was conducted as per ISO 23317 before and after soaking in Hank's solution. Microscopy analysis showed a continuous composition gradient with alternate layers of hydroxyapatite and titania. The in-vitro results which showed biological response of FGC revealed bone-like apatite growth after being immersed in Hank's solution for 7–14 days. Mechanical behavior evaluated via fracture toughness and interfacial adhesion tests showed good stability after 14 days' immersion in Hank's solution which was an encouraging result from a long-term mechanical stability point of view when compared to that of monolayer coating structure.

4.3.1.3 Suspension Plasma Spraying (SPS)

This novel spraying technique has a distinct advantage over other thermal spray techniques, that is the capability of depositing nanometer-range powder particles. Basically, this is a modified version of the APS technique in which the suspension of solid particles is used to form coatings. Large suspension droplets consisting of solid phase tend to undergo fragmentation when fed into the plasma jet leading to the formation of submicron and nanosize aggregates. The mechanism of smaller droplets due to aerodynamic breakdown or secondary fragmentation is shown in Figure 4.6 [32]. The feedstock of this technique can cover most of the materials including alloys, ceramics, and metal oxides. By maintaining a feeding rate of 5 to 100 ml/min one can obtain deposition efficiency in the range of 50% to 70%. The usage of fine particles helps in easy melting and helps in the reduction of pore size and decreases the tendency of micro-crack formation.

Cattini et al. [33] studied the mechanical behavior and suitability for biomedical application by conducting in-vitro tests on bioactive glass/hydroxyapatite FGC developed using SPS. Feedstock suspension of 10% solid phase and 90% ethanol was used to develop a total thickness of 80 μm. Microstructure analysis showed the formation of compact, continuous, and compositional gradient coatings. The Martens hardness and elastic modulus of graded structure were 2.42 GPa and 3.20 GPa respectively which was comparatively higher than observed for an area which was closer to the surface. In-vitro tests showed that the developed FGC was suitable for biomedical applications due to the presence of bioactive glass which exhibited apatite-forming ability in simulated body fluid. Functionally graded bioceramic coatings with a combination of fluoridated hydroxyapatite/calcium silicate developed using SPS (Model MC 60, Medicoat AG, Switzerland) were reported by Yin et al. [34]. The feedstock suspension consisting of 83.5% deionized water, 15% solid phase, and 1.5% ammonium polyacrylate was employed for the development of FGC. The spray parameters of feeding rate of 40 mL/min, spraying distance of 38 mm, and torch linear velocity of 700 mm/s were chosen. Assessment of both the ability to form apatite in vitro and antibacterial activity was conducted at 37°C under static and aerobic conditions. Microstructure analysis showed splats, rough and porous surface structure, while cross-section analysis showed a lamellar structure with no cracks at the interface between the substrate and FGC. In-vitro tests revealed apatite-forming ability of the FGC, recommending it as a potential candidate material for bone implants. Antibacterial tests showed that FGCs were capable of inhibiting bacterial adhesion and immobilizing certain bacteria. Apart from biomedical applications, the SPS technique was also used for the development of functionally graded $La_2Zr_2O_7$/8YSZ thermal barrier coating materials [35]. Feedstock suspensions consisting of 69% ethanol, 30% solid phase, and 1% polyethylene glycol were used. Using a set up made

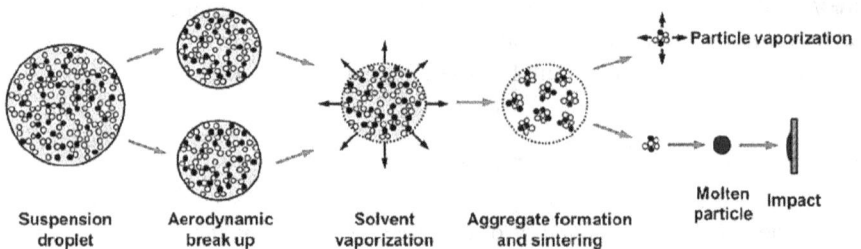

FIGURE 4.6 Mechanism of submicron or nanosize aggregate formation in SPS technique [32].

by Metco (Sulzer Metco Company, Switzerland), the coatings were deposited at a 25 mL/min feed rate and spraying standoff of 50 mm. Surface microstructure analysis showed good melting with efficient spreading of melted particles. Cross-sectional analysis revealed the successful formation of graded structure with a gradual change in composition. Thermal cycling tests conducted at 1000 and 1200°C showed that the FGC showed better performance and prolonged life when compared with single and double ceramic layer coatings. The improved thermal cycling life of the FGC was attributed to the formation of gradual variation in composition along the thickness direction.

4.3.1.4   Vacuum Plasma Spraying (VPS)

This technique is quite different from conventional thermal spray techniques as it employs a high-velocity and high-power plasma gun which operates in a low-pressure or partial-vacuum environment. VPS was developed with the primary intention of improving the quality, durability, and reliability of coatings by depositing in a low-pressure noble gas atmosphere. The spray chamber which is filled with inert gas is maintained at a pressure as low as ~100 mbar to avoid oxidation. The reduced pressure here helps in achieving high gas as well as particle velocities which are capable of depositing molten particles of size in the range of 10 to 100 μm. The spraying speed is much faster than in air due to which the coatings produced are oxide free, high density, and possess higher bonding strength. VPS coatings are capable of enhancing abrasion, friction, erosion, and cavitation of substrate material [35, 36].

For heat shield applications the coating should sustain temperatures as high as 1000 K without much significant change in structure. Choi et al. [37] reported the development of Ni-YSZ FGC with varying YSZ content from 20% to 80% and NiCrAlY as the metallic bond coating. A high-temperature plasma torch was used to conduct heat shield performance tests where the temperature of the torch was set to 2000 K. One-dimensional steady-state heat transfer expression was used to calculate the temperature at the interface between the substrate and bond coating. The presence of 100% YSZ coating as a top coat helped in decreasing the temperature drastically while in FGC the drop was quite gradual. The FGC layers played a role of buffer material in minimizing the thermal stress between the top YSZ layer and the bottom bond coat layer. For applications like nuclear fusion reactors the interaction of structural materials like reduced activation ferritic martensitic steel and plasma leads to lowering of design life of steel. In such cases many have opted to provide tungsten coating which however leads to the premature failure of coating due to large residual stresses. Keeping this in mind Vaßen et al. [38] reported the development of tungsten/EUROFER97 FGC using the VPS technique. The alternate layers of steel and tungsten were coated onto a EUROFER97 substrate using a VPS system (make: Oerlikon Metco, Wohlen Switzerland). Residual stress measurements were conducted by increment drilling of holes using TiN coated endmills. The results showed a reduction in localized deformation and degradation of coatings at the interface region due to the right approach adopted for the graduation of the interface. The VPS technique was extended to biomedical applications in order to develop biocompatible and chemically stable FGCs. Kuo et al. [39] studied the biological properties of functionally graded tantalum layers deposited on Ti6Al4V alloy. Spray parameters like powder feed rate of 40 g/min, spray distance of 250 mm, and chamber pressure of 150 mbar were used for depositing coating using a VPS system (Metco A-3000 VPS machine, Sulzer Metco, Switzerland). Surface modification was carried out by employing a combination of alkali and heat treatment processes. A bone-like apatite test was conducted in simulated body fluid (SBF) at a temperature of 37°C for 3–14 days. Observation after 3 days revealed dense apatite formation on alkali and heat-treated FGC while the tantalum coating or Ti6Al4V alloy showed no such signs. The increased bioactivity was due to the formation of amorphous sodium tantalite during alkali treatment. Overall these studies showed that the VPS technique can be used for the development of FGC for almost all kind of applications.

4.3.2 Physical Vapor Deposition

The approach in which the solid source is evaporated, the formation and transportation of vapor phase to substrate, and the condensation of vapor on the substrate surface leading to the formation of thin films is called physical vapor deposition (PVD). The process for the development of FGM has a small modification; instead of one the process employs two solid sources containing the material to be coated. The process of atomization depends on the type of PVD process that is employed. These PVD processes are sputtering facilitated PVD, plasma spray facilitated PVD, and evaporation facilitated PVD. The basic principle of operation of sputtering facilitated PVD is that sputtering gas ions are targeted towards the target material leading to the ejection of atoms from its surface. The different types of power source are direct current, magnetron sputtering, radio frequency sputtering, and reactive sputtering. The plasma spray facilitated PVD is also a novel deposition technique developed by Sulzer-Metco. Employing an operating pressure of up to 2 mbar and high plasma stream velocity of 2000 m/s facilitates easy vaporization of the feedstock material. Finally in evaporation facilitated PVD the material to be deposited is first melted and then vaporized using an appropriate heat source. Based on the heat source this process is further classified into ion beam-assisted evaporation and molecular beam epitaxy. Two important classes of PVD, namely, pulse laser deposition and electron beam PVD, used for the development of FGC, are presented in upcoming sub-sections.

4.3.2.1 Electron Beam Physical Vapor Deposition (EB-PVD)

The electron beam generated from the electron source is irradiated onto the materials that have to be deposited. Due to the irradiation of the electron beam the material is evaporated and deposited onto the surface of the substrate. This process is generally employed to evaporate materials which are otherwise difficult to evaporate using standard resistive thermal evaporation techniques. EB-PVD coatings are generally applied on turbine aerofoil components like blades and vanes to sustain high thermo-mechanical strains. The advantage of employing EB-PVD is that refractory metals like titanium boride, zirconium boride, and titanium carbide can be evaporated without undergoing decomposition. The deposition rates are very high due to which they are preferred for producing thermal barrier and wear-resistant coatings.

Movchan and Rudoy [40] reported the development of gradient thermal barrier coatings using the EB-PVD process. The fabrication process included pressing a multicomponent tablet $(Al–Al_2O_3–ZrO_2(Y_2O_3))$ against a ceramic ingot $(ZrO_2–7\%Y_2O_3)$ followed by evaporation of the tablet first and then the ceramic ingot next. Prior to that, an NiCrAlY bond coat of 50 µm thickness was deposited from a separate source onto the Inconel 718 substrate. The deposition rates for the bond coat and FGC were ~25 µm and ~10 µm respectively. The FGC coating had a thickness of 150 µm with a top layer of $ZrO_2–7\%Y_2O_3$, a transition gradient zone consisting of Al, Al_2O_3, and ZrO_2, and the NiCrAlY bond coat. Microstructure analysis showed smooth transition in composition from the bond coat to the transition gradient zone with no structural imperfections. The microhardness measurements measured from the bond coat to the transition gradient zone showed values varying from 508 VHN to 729 VHN. In another work, Meng et al. [41] reported functionally graded Ni–YSZ anode coatings with gradient dispersion of NiO. EB–PVD equipment with a deposition rate of 0.6 µm/min was used to evaporate two ingots of NiO and YSZ respectively. The SUS430 steel was used as the substrate material but prior to deposition the substrate was cleaned using supersonic waves in ethanol. The other parameters were as follows: distance between ingot and substrate of 300 mm, substrate temperature of 650°C, and pressure of deposition chamber maintained at 1×10^{-2} Pa. The final Ni–YSZ coating was obtained by subjecting NiO–YSZ to reduction under hydrogen atmosphere maintained at 800°C. Unlike in other EB–PVD coatings here no columnar structure was formed and a high porosity level of 33% was achieved. Such FGC suits well for applications like anode coating material for solid oxide fuel cells owing to its low manufacturing cost, good catalytic activity, and long-term stability.

4.3.2.2 Pulsed Laser Deposition (PLD)

This an important process of PVD for thin film deposition using a high-power laser beam focused on the target material. Due to the laser beam the surface of the target material is vaporized and condensed vapors are deposited onto the surface of the substrate. The PLD set-up consists of a vacuum chamber, pulsed laser, and rotating target holder. The electrical field generated is strong enough to eject the electrons from the substrate surface and this occurs within 10 picoseconds of a nanosecond laser pulse. The PLD process can be modified into plasma-assisted PLD by enhancing the interaction between vapors and ambient gas by using plasma. If an auxiliary ion beam is used to irradiate the substrate surface then such process is known as ion beam-assisted PLD [42].

Tanaskovic et al. [43] reported the development of bioactive glass-apatite FGC on titanium substrate using PLD. Sintered hydroxyapatite and bioglasses with 57 SiO_2 content were selected as targets and UV KrF excimer laser source was used for the ablation of targets. The deposition was carried out when the substrate was maintained at a temperature of 400°C and the laser beam was targeted at 45°. In order to improve morphology and avoid piercing of the deposited coating the targets were rotated continuously. It was ensured that the glass layer came first into contact with

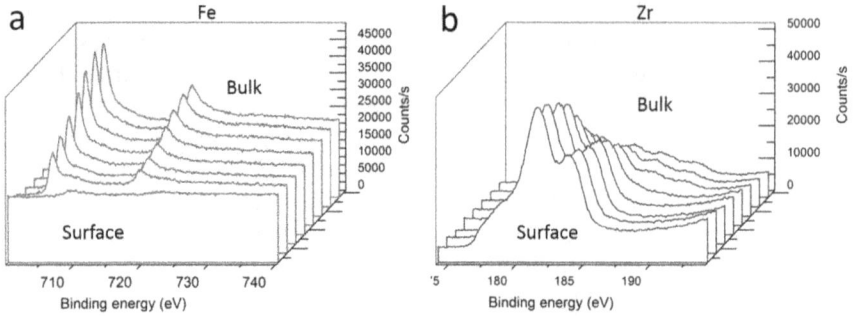

FIGURE 4.7 XPS measurements on graded coatings depicting (a) Fe(2p) shown in red lines and (b) Zr(3d) shown in blue lines are the intensity peaks right from the substrate/coating interface to the outermost layer [44].

the Ti substrate due to the similar thermal expansion coefficients of both materials. Microscopy analysis showed that the deposited layer films consisted of spherical droplets due to laser ablation of the brittle natured targets. Despite the presence of defects, the FGC showed good adherence to the titanium substrate as revealed by scotch and scratch tests. Such composition and mechanical stability ensures a suitable environment for the development of cell culture. The PLD process was employed for the development of yttria-stabilized zirconia FGC for thermal barrier applications [44]. A 355 nm, 6 W picosecond laser (Make: Coherent Talisker 355-4) as energy source, SUS 316L and 8% YSZ as target materials, and two molecular pumps capable of maintaining $\sim 10^{-5}$ Torr pressure inside the vacuum chamber were used for coating deposition. X-ray photoelectron spectroscopy (XPS) depth profile measurements showed that the intensity peak for Fe(2p) decreased gradually from substrate/coating to outer layer of coating while that of Zr(3d) showed the reverse trend which is quite clearly shown in Figure 4.7. The adhesion strength measurement conducted using the Rockwell indentation cracking method showed good mechanical stability and adhesion strength. A further thermal conductivity test was conducted for which a 1070 nm multi-mode fiber laser (Make: IPG YLS-2000) and power of 100 W were used as heat source. The thermal conductivity results showed a higher value of 16.01 W/mK for the pure steel sample, 4.86 W/mK for FGC, and a very low value of 1.14 W/mK for YSZ. So it is quite clear that most of the ceramics having low thermal conductivity values are the most promising candidate materials for thermal barrier applications.

4.3.3 CHEMICAL VAPOR DEPOSITION

In this process the substrate maintained at higher temperatures is exposed to the volatile precursors which either decompose or react with the substrate to form the desired thin film on its surface. The main point to be noted is here that the chemical reactions occur only near and on the surfaces which are maintained at elevated temperatures. The growth rate of mullite FGC with respect to the deposition temperature and total pressure can be seen in Figure 4.8 [45]. It is observed that at low metal chloride partial pressures the amorphous aluminosilicates grow very fast and a further increase in pressure leads to the formation of crystalline mullite as shown in Figure 4.8a and b. Further increases in temperature and metal chloride partial pressure lead to increases in grain size and further increases in these parameters resulted in a cauliflower-like structure as seen in Figure 4.8c and d. One can deposit a wide variety of materials including metals, ceramics, oxides, nitrides, carbides, and intermetallic compounds. The distinct advantage of CVD over PVD is that it can cover the complete surface of the substrate to be coated including the restricted surfaces as well whereas PVD needs line-in-sight between source and surface to be coated. Further, there is no need for high vacuum like the PVD process, and one can achieve high deposition rates. This process is capable of changing various attributes like the mechanical, optical, and electrical properties of the substrate material. Because of this many industries use CVD coatings to provide protection against erosion, corrosion, and high temperatures. There are different types of CVD process and based on the chemical reactions initiated they can be differentiated from one another. If classification is done from an operating pressure point of view then derivatives of CVD are atmospheric pressure CVD, low-pressure CVD, and ultrahigh vacuum CVD. Let's see a few important derivates of the CVD process, plasma-assisted/enhanced CVD and metal-organic CVD used in the development of FGC. Most of these processes are targeted for wear-resistant and biomedical applications using combinations of new materials.

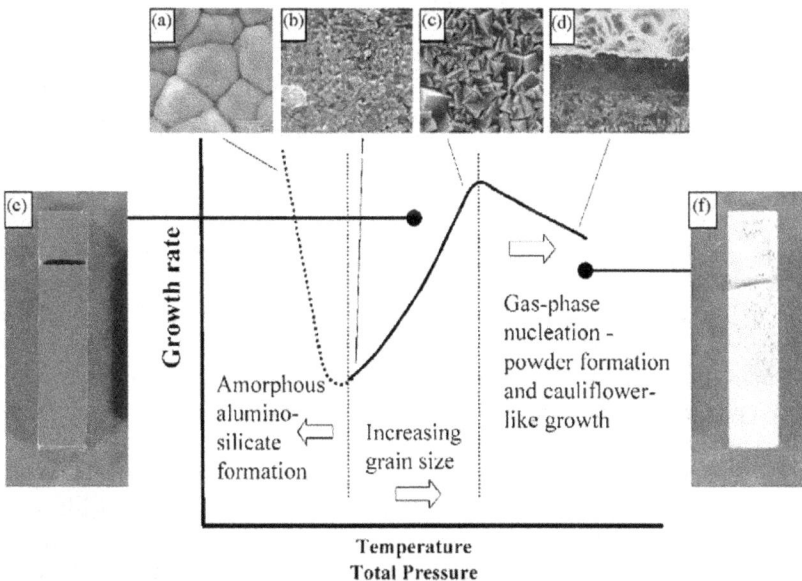

FIGURE 4.8 Schematic showing the growth rate of mullite coatings with increasing deposition temperature and pressure [45].

4.3.3.1 Plasma-Enhanced/Assisted CVD

This process is used to develop thin films and is accepted widely because it can operate at low temperatures unlike other CVD processes which are thermally driven. The process includes the ignition of plasma consisting of ionized gas and electrons, inelastic collision of gas molecules and electrons, formation of reactive species like free radicals, and deposition of thin films with the aid of reactive species. Abedi et al. [46] reported the development of compositionally graded TiSiCN coatings on a substrate made up of H13 hot-working tool steel using the plasma-enhanced CVD process. To start with the titanium was coated by introducing argon, hydrogen, and $TiCl_4$ vapors into the vacuum chamber. The nitrogen and methane were injected into the vacuum chamber at constant flow rates of 30 and 10 standard cubic centimeters per minute. The process of deposition continued for about 3 hours with the temperature of the substrate maintained at 500°C. The hardness and elastic modulus were found to be 41 GPa and 341 GPa respectively. The wear test conducted using a pin-on-disc tribometer revealed that the normal force required in case of FGC for initiation of crack and delamination was 117% and 40% higher than that of a single layer of similar coating. Such highly wear-resistant materials are usually utilized in the cutting tool industry. In another work Damerchi et al. [47] reported the development of nanostructured TiCN FGC using plasma-assisted CVD process on plasma nitrided H13 steel towards punching tool applications. Prior to coating deposition the steel was austenitized at 1050°C for 45 minutes followed by oil quenching to ambient temperature. Deposition was carried out using the PACVD system (make: Plasma Fanavar Amin (PFA) Co.) having an 800 W DC power supply. TiCN FGC was deposited at three different temperatures, 450, 475, and 500°C, with hydrogen, argon, and nitrogen flow rates of 200, 100, and 40 standard cubic centimeters per minute. A deposition temperature of 475°C resulted in a high hardness value of 30 GPa and the lowest wear volume of 7.4×10^6 μm^3. The wear profile of FGC deposited at 475°C showed a lower depth of abrasion marks when compared to other coatings.

4.3.3.2 Metal-Organic CVD

This CVD process employs organo-metallic precursors for the depositing of high-purity crystalline micro- and nanometer thin films. The precursors are usually transported to the substrate surface with the help of carrier gases like argon and nitrogen. Initially this method was employed for the deposition of epitaxial layers of semiconductor materials. High deposition rates of 100 nm/min can be achieved with controllable coating stoichiometry. The advantages of this process include the deposition of conformable coatings, deposition on large areas, proven capability to produce multilayers, and even composition graded layers. Due to its high flexibility, the desired composition on suitable substrate material can be deposited at lower cost. Sato et al. [48] developed the Ca–Ti–O/Ca–P–O functionally graded films using the metal-organic CVD process for possible application in medical implants. The precursors, $Ca(dpm)_2$, $Ti(O-i-Pr)_2(dpm)_2$, and $(C_6H_5O)_3PO$, were heated and carried into CVD reactors using inert gas like argon. The substrate temperature varied from 600 to 800°C, and the precursor vapors mixed with oxygen were deposited onto the substrate. An immersion test into Hank's solution was conducted on Ca–Ti–O/ Ca–P–O films to study the apatite layer formation. The deposition rate relationship with substrate temperature was obtained in Arrhenius format and found that the increase in substrate temperature led to an increase in deposition rate. The apatite formation rate showed dependency on the surface morphology of the calcium titanate film. Further apatite was formed on the surface after hours of immersion and tended to increase with the increase in immersion period. Shiraishi et al. [49] reported the development of calcium titanate functionally graded films on commercial-grade titanium substrate. The CVD pressure was kept at 0.8 kPa, and the substrate temperature was maintained at 800°C. The level of osseointegration was evaluated using push-in tests where harvested femurs consisting of implants were incorporated into autopolymerizing resin. In general, a loading rate of 1 mm/min is chosen for push-in tests and the values obtained from the test define the breakpoint load at the tissue/implant interface. The push-in value of CTO 800 obtained after 2 weeks was significantly higher than titanium indicating the capability of calcium titanate FGC in promoting apatite nucleation and accelerating the process of osseointegration. These results supported the use of calcium titanate as a potential biomaterial for medical devices.

4.4 COMPUTATIONAL MODELING AND ANALYSIS

Realistic predictions of the mechanical and thermal behavior of FGC need constitutive relations along with corresponding physical and mechanical property data. Most importantly the property data largely depend upon the properties of individual phases and other details like volume fraction, orientation, and spatial distribution. Due to the cost and time factor involved in experimentation, various approaches like rules of mixture, variational principles of thermo-mechanics, a micromechanics approach, an empirical approach, and fuzzy logic techniques are employed. However, in present days, material design and analysis using computers and robust software have become very important tools which are not only informative but also save time and decrease cost. Commercial players always value saving time and cost which is spent on experimentation and testing. Many approaches have already been proposed to study the effect of mechanical and thermal loads on FGM. These approaches help a design engineer to arrive at an appropriate compositional gradient, improve the mechanical and thermal behavior, and most importantly improve the lifetime of components made up of FGM. Take the example of thermal barrier coatings which are commonly developed using the thermal spray process. On cooling, due to the difference in the coefficient of thermal expansion of the substrate and coating, thermal residual stresses develop. These residual stresses lead to premature failure of the coating which eventually affects the component life. Instead of conventional thermal barrier coatings, if FGC are used in place, then it is very important to estimate the thermal stresses and their distribution in the coatings. However one should understand that for simple geometries and material properties the analytical solutions can be used but in case of complex geometries numerical solutions like the finite element method can be used. In a nutshell the computational modeling and analysis of FGC are quite different from conventional coatings and thin films in terms of equilibrium equations, boundary conditions, constitutive relations, and homogenization.

Measuring the mechanical properties of each coating layer in FGC using a conventional mechanical test like the tension test is quite impossible until the thickness is around several hundred microns. In such cases nanoindentation can be used to measure the hardness and elastic modulus of FGC. Taking a cue from this, Kang et al. [50] reported the application of elastic-plastic finite element analysis to determine the stress distribution and deformation under different loading conditions. The yield strength of FGC was determined by combining nanoindentation and the finite element analysis model. The nanoindentation experiments were carried out on $Ti_{1-x}C_x$/DLC FGC deposited on 440C steel using a hybrid combination of pulsed laser deposition and magnetron sputtering. Based on measured nanoindentation data, the yield strength of FGC was calculated from the finite element model developed for nanoindentation simulation. The von Mises stress profile ($\delta = 200$ nm and $\mu = 0.0$) obtained inside the coating system revealed moderate stress variation across the thickness for Ti/TiC/DLC FGC when compared to that of monolayered diamond-like carbon coating (DLC). This was attributed to gradual change in mechanical properties for Ti/TiC/DLC FGC across the thickness. In a similar note another interesting work was reported by Joshi and Ng [51] on optimizing the composition of nickel–zirconia FGC to obtain low and uniform stress field when subjected to thermal loading. The stress resulting from thermal loading and its distribution in FGC were analyzed using finite element software named LUSAS (version 13). The nodal temperature of the top surface of the FGC was maintained at 500 K while the substrate bottom was kept at 300 K. The effective properties were described by polynomial functions against the volume percentage of ZrO_2 content, and using effective properties different FGC distribution profiles were attempted. Among various distribution profiles obtained for thermal stress, the quadratic profile showed the largest drop in thermal stress. In addition to this the configuration where the profile of FGC follows the concave power law showed lower peak stress, especially von Mises stress, at interfaces and the substrate was lowest. When stiff coatings are deposited on the softer substrate then delamination of coatings takes place due to buckling instabilities. In general two mechanisms that cause coating failure or interface delamination are wrinkles and buckling. Keeping this in mind, Reinoso et al. [52] reported investigation of the delamination phenomenon in FGC from the elastic substrates. For detailed literature review, the authors finalized on the idea to combine nonlinear fracture mechanics and solid shell finite element model for thin film to develop a 3D nonlinear computational framework. The analysis started with an introduction of mechanical models for cohesive interface and functionally graded thin film. Interface element formulation with large displacement was proposed to simulate nonlinear fracture at the interface between the metallic bond coat and the ceramic top coat. The results obtained from the computations were compared with benchmarked 2D results available in the literature and found to be accurate. Most of these studies were able to predict the stress distribution and optimal composition gradient that were capable of affecting the coating quality, integrity, and performance of FGC.

4.5 APPLICATIONS

Recent development in technologies like gas turbines, orthopedic implants, cutting tools, energy production, punching tools, nuclear fusion, and ceramic engines are dictating the need for novel and multifunctional materials which are capable of meeting the needs of the above applications. Each industry has its own criteria which have to be met by the FGC to be used in appropriate application. For instance, reliability and durability are key issues for space vehicle components and nuclear fusion applications while the cost factor doesn't have much significance. On the other hand the key issue for engine components and cutting tools is the ratio of cost by performance. Using the FGM approach, functionally graded coatings were developed for various applications like thermal barrier coating, biologic coatings for bone and dental implants, and corrosion-, wear-, cavitation-, and erosion-resistant coatings. Many research works have resulted in positive outcomes on the laboratory scale while some of them are in practical use already. In this section the present and possible applications of FGCs in different disciplines of engineering are provided in brief.

4.5.1 FUNCTIONALLY GRADED THERMAL BARRIER COATINGS

A large number of functionally graded coatings have been suggested as thermal barrier coatings for internal combustion engines, and commercial and military gas turbine applications. Figure 4.9 shows a cross-section of an industrial gas turbine

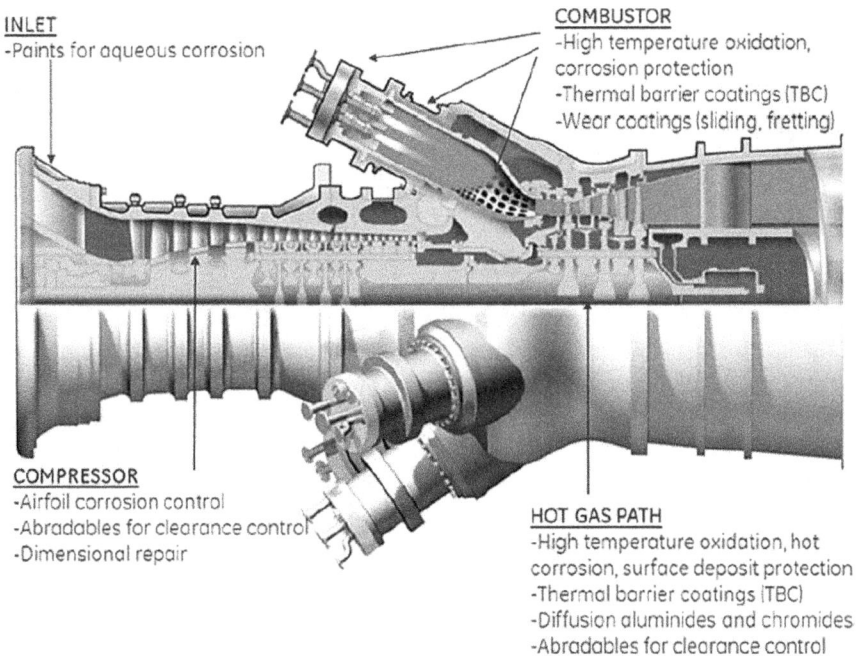

INLET
-Paints for aqueous corrosion

COMBUSTOR
-High temperature oxidation, corrosion protection
-Thermal barrier coatings (TBC)
-Wear coatings (sliding, fretting)

COMPRESSOR
-Airfoil corrosion control
-Abradables for clearance control
-Dimensional repair

HOT GAS PATH
-High temperature oxidation, hot corrosion, surface deposit protection
-Thermal barrier coatings (TBC)
-Diffusion aluminides and chromides
-Abradables for clearance control

FIGURE 4.9   Various components of gas turbine engine and potential locations where conventional coatings are used and can be replaced by FGC [53].

engine and various components where conventional thermal barrier coatings are employed [53]. The conventional coatings can be replaced by FGC in these components for better performance and improved life. By providing thermal insulation to metallic components from combustion gas stream one can minimize or eliminate thermal fatigue, hot corrosion, wear, and erosion, thereby extending the life of components [54, 55]. Strictly from a material combination point of view, YSZ is used at different locations as top coat, as bottom layer, and as intermediate layer in different FGC due to its low thermal conductivity and coefficient of thermal expansion. The oxides like ZrO_2, Al_2O_3, and TiO_2 are added as one of the constituents in FGC for pistons in internal combustion engine applications. The components where FGC can find potential application as a thermal barrier are cylinder heads, valves, piston crowns, and rings. Complex oxides, perovskites, and hexa-aluminates which are compatible with YSZ demonstrated good thermal insulation properties and a long thermal cycling lifetime which are usually preferred for burner rig applications and engine components like exhaust manifold and pistons [53, 56, 57]. FGC is a potential candidate material for extreme conditions like space shuttle applications because the temperature difference between inside and outside is over 1000°C and appropriate compositional design can serve this purpose. Conventional coatings tried for plasma-facing components suffer from premature failure due to stress at the interface and high temperatures that cross the tolerable level due to high reactor load. In such applications suitable composition gradient thermal barrier coatings based on tungsten and steel can be tried.

4.5.2 FUNCTIONALLY GRADED BIOMEDICAL/BIOACTIVE COATINGS

Biologic coatings on bone and dental implants need to have multifunctional properties and for this many FGCs have been suggested. In this regard hydroxyapatite-based FGCs are usually preferred to promote osseointegration owing to their crystallographic similarities with the inorganic constituents of bone. Hydroxyapatite or calcium phosphate-based FGC are tried in the field of dentistry and bone regeneration. Although calcium phosphate was used potentially for bone grafts or bone substitute in the early 1920s in the form of cements and paste, the momentum of development of FGC using calcium phosphate has gained pace recently. This is mainly because of the latest developments in the field of gaseous phase deposition techniques, and the compositional gradient in these coatings provides good mechanical strength and gradient in bioactivity. Such biologic FGCs are produced using plasma spraying, suspension plasma spraying, DC magnetron sputtering, RF magnetron sputtering, and pulsed-laser deposition. Further titanium is used as an implant material due to its load-bearing capability but from a compatibility point of view it is rejected by the human body leading to the release of toxic metallic ions. Due to this, adverse medical conditions can occur and lead to allergy and inflammation in the human body. In such cases functionally graded coatings of bioactive ceramics based on calcium phosphate have been proposed. Different ceramics like Al_2O_3, YSZ, TiO_2, and bioactive glass are other constituents of FGC used along with hydroxyapatite for mechanical stability and crack prevention.

4.5.3 Functionally Graded Coatings for Cutting Tools

Although the optimization of cutting parameters can improve the machining efficiency, improvement might not be significant enough. Modern cutting tools need to have high wear resistance, high fracture toughness, and crack-resistant properties. If cutting tool doesn't have such multifunctional properties than they can be achieved by applying FGC on their surface. Functionally graded coating and substrate materials such as $TiCN/Al_2O_3$–TiN and WC–Ti(C,N)–(Ta,Nb)C–Co were proposed to replace conventional tool materials. Functionally graded cemented carbide coating is used for making Tiebar-cut punch tools which are generally used for cutting lead frame. CVD processed TiN–HT–Ti(C,N)-κ-Al_2O_3–TiN functionally graded outer layers were used to improve the wear performance of cutting tool inserts [58, 59].

4.6 CONCLUSIONS

The evolution of functionally graded materials started with the application in the area of thermal barrier to reduce thermal stress when exposed to high temperatures. Due to the gradient in composition and structure the functionally graded materials can exhibit incompatible functions like wear, erosion, cavitation, corrosion, heat, and oxidation resistance. Conventional gaseous phase deposition techniques are used to deposit the functionally graded coatings without much modification in the deposition setup. The properties of the functionally graded coating can be tailored to the needs of application right from heat shield to biomaterials. This chapter provides a basic introduction and research studies conducted by various researchers on various gaseous phase processing techniques of functionally graded materials. Further, it will save time for academicians and researchers in researching gaseous phase deposition techniques and the latest developments in functionally graded coatings.

REFERENCES

1. Mannan, S., Parameswaran, V., Basu, S., Stiffness and toughness gradation of bamboo from a damage tolerance perspective. *International Journal of Solids and Structures*, 2018, 143: pp. 274–286.
2. Rossetti, L., et al., The microstructure and micromechanics of the tendon-bone insertion. *Nature Materials*, 2017, 16: pp. 664–670.
3. Lui, P.P., Zhang, P., Chan, K.M., Qin, L., Biology and augmentation of tendon-bone insertion repair. *Journal of Orthopaedic Surgery and Research*, 2010, 5: pp. 59.
4. Miyamoto, Y., Kaysser, W., Rabin, B.H., Kawasaki, A., Ford, F.G., *Functionally Graded Materials: Design, Processing, and Applications*, 1999, Springer, New York.
5. Heimann, R.B., Kleiman, J.I., Litovsky, E., Marx, S.N.R., Petrov, S., Shagalov, M., Sodhi, R.N.S., Tang, A., High-pressure cold gas dynamic (CGD)-sprayed alumina-reinforced aluminum coatings for potential application as space construction material. *Surface and Coatings Technology*, 2014, 252: pp. 113–119.
6. Shankar, A.R., Mudali, U.K., Preparation of the plasma sprayed graded thermal barrier coating by co-injection of premixed powders through a single plasma torch. *Surface Engineering*, 2018, 34:728–736.

7. Roy, S., Functionally graded coatings on biomaterials: A critical review. *Materials Today Chemistry*, 2020, 18: pp. 100375.

8. Wu, H., Yu, J., Song, W., Zou, J., Song, Q., Zhou, L., A critical state-of-the-art review of durability and functionality of open-graded friction course mixtures. *Construction and Building Materials*, 2020, 237: pp. 117759.

9. Bajaj, K., Shrivastava, Y., Dhoke, P., Experimental study of functionally graded beam with fly ash. *Journal of the Institution of Engineers: Series A*, 2013, 94: pp. 219–227.

10. Torelli, G., Fernandez, M.G., Lees, J.M., Functionally graded concrete: Design objectives, production techniques and analysis methods for layered and continuously graded elements. *Construction and Building Materials*, 2020, 242: pp. 118040.

11. Smith, R.W., Knight, R., Thermal spraying II: Recent advances in thermal spray forming. *JOM*, 1996, pp. 16–19.

12. Sampath, S., Herman, H., Shimoda, N., Saito, T., Thermal spray processing of FGMs. *MRS Bulletin*, 1995, pp. 27–31.

13. Latka, L., Pawlowski, L., Winnicki, M., Sokolowski, P., Malachowska, A., Kozerski, S., Review of functionally graded thermal sprayed coatings. *Applied Sciences*, 2020, 10: pp. 5153.

14. Matejicek, J., Tungsten-steel composites and FGMs prepared by argon-shrouded plasma spraying. *Surface and Coatings Technology*, 2021, 406: pp. 126746.

15. Singh, J., Chatha, S.S., Singh, H., Synthesis and characterization of plasma sprayed functional gradient bioceramic coating for medical implant applications. *Ceramics International*, 2020, https://doi.org/10.1016/j.ceramint.2020.12.039.

16. Taleghani, P.R., Valefi, Z., Ehsani, N., Evaluation of oxidation and thermal insulation capability of nanostructured $La_2(Zr_{0.7}Ce_{0.3})_2O_7$/YSZ functionally graded coatings. *Ceramics International*, 2020, https://doi.org/10.1016/j.ceramint.2020.12.012.

17. Torabi, S., Valefi, Z., Ehsani, N., Ablation behavior of SiC/ZrB_2 multilayer coating prepared by plasma spray method. *Metallurgical and Materials Transactions A*, 2020, 51: pp. 1304–1319.

18. Mendoza, M.Z., Rueda, C.J.M., Sanchez, A.M., Lopez, F.J., Thermal cyclic oxidation of NiCoCrAlYTa coatings manufactured by combustion flame spray. *Materials Today Communications*, 2020, 25: pp. 101617.

19. Emmerich, T., Vaßen, R., Aktaa, J., Thermal fatigue behavior of functionally graded W/EUROFER-layer systems using a new test apparatus. *Fusion Engineering and Design*, 2020, 154: pp. 111550.

20. Shreeram, B., Rajendran, I., Kumar, E.R., Tailoring of functionally graded mullite: La_2O_3 coatings by transferred arc plasma for thermal barrier coatings. *Journal of Inorganic and Organometallic Polymers and Materials*, 2018, 28: pp. 2484–2493.

21. Henao, J., et al., HVOF hydroxyapatite/titania-graded coatings: Microstructural, mechanical, and in vitro characterization. *Journal of Thermal Spray Technology*, 2018, 27: pp. 1302–1321.

22. Vakilifard, H., Ghasemi, R., Rahimipour, M., Hot corrosion behaviour of plasma-sprayed functionally graded thermal barrier coatings in the presence of $Na_2SO_4 + V_2O_5$ molten salt. *Surface and Coatings Technology*, 2017, 32: pp. 238–246.

23. Carpio, P., Salvador, M.D., Borrell, A., Sánchez, E., Thermal behaviour of multilayer and functionally-graded $YSZ/Gd_2Zr_2O_7$ coatings. *Ceramics International*, 2017, 43: pp. 4048–4054.

24. Xia, L., Xie, Y., Fang, B., Wang, X., Lin, K., In situ modulation of crystallinity and nano-structures to enhance the stability and osseointegration of hydroxyapatite coatings on Ti-6Al-4V implants. *Chemical Engineering Journal*, 2018, 347: pp. 711–720.

25. Rezapoor, M., Razavi, M., Zakeri, M., Rahimipour, M.R., Nikzad, L., Fabrication of functionally graded Fe-TiC wear resistant coating on CK45 steel substrate by plasma spray and evaluation of mechanical properties. *Ceramics International*, 2018, 44: pp. 22378–22386.

26. Liang, P., et al., Duplex and functionally graded Al@NiCr/8YSZ thermal barrier coatings on aluminum substrates. *Journal of Alloys and Compounds*, 2019, 790: pp. 928–940.

27. Saremi, M., Valefi, Z., Thermal and mechanical properties of nano-YSZ-Alumina functionally graded coatings deposited by nano-agglomerated powder plasma spraying. *Ceramics International*, 2014, 40: pp. 13453–13459.

28. Jokanovic, V., Vilotijevic, M., Colovic, B., Jenko, M., Anzel, I., Rudolf, R., Enhanced adhesion properties, structure and sintering mechanism of hydroxyapatite coatings obtained by plasma jet deposition. *Plasma Chemistry and Plasma Processing*, 2015, 35: pp. 1–19.

29. Cannillo, V., Lusvarghi, L., Sola, A., Production and characterization of plasma-sprayed TiO_2-hydroxyapatite functionally graded coatings. *Journal of the European Ceramic Society*, 2008, 28: pp. 2161–2169.

30. Ivosevic, M., Knight, R., Kalidindi, S.R., Palmese, G.R., Sutter, J.K., Solid particle erosion resistance of thermally sprayed functionally graded coatings for polymer matrix composites. *Surface and Coatings Technology*, 2006, 200: pp. 5145–5151.

31. Stewart, S., Ahmed, R., Itsukaichi, T., Contact fatigue failure evaluation of post-treated WC-NiCrBSi functionally graded thermal spray coatings. *Wear*, 2004, 257: pp. 962–983.

32. Gan, J.A., Berndt, C.C., Nanocomposite coatings: Thermal spray processing, microstructure and performance. *International Materials Reviews*, 2015, 60: pp. 195–244.

33. Cattini, A., Bellucci, D., Sola, A., Pawłowski, L., Cannillo, V., Suspension plasma spraying of optimised functionally graded coatings of bioactive glass/hydroxyapatite. *Surface and Coatings Technology*, 2013, 23: pp. 118–12.

34. Yin, X., Bai, Y., Zhou, S., Ma, W., Bai, X., Chen, W., Solubility, mechanical and biological properties of fluoridated hydroxyapatite/calcium silicate gradient coatings for orthopedic and dental applications. *Journal of Thermal Spray Technology*, 2020, 29: pp. 471–488.

35. Wang, C., et al., Optimized functionally graded $La_2Zr_2O_7$/8YSZ thermal barrier coatings fabricated by suspension plasma spraying. *Journal of Alloys and Compounds*, 2015, 49: pp. 1182–190.

36. Azarmi, F., Vacuum plasma spraying. *Advanced Materials and Processes*, August 2005: pp. 37–39.

37. Choi, K.H., et al., High-temperature thermo-mechanical behavior of functionally graded materials produced by plasma sprayed coating: Experimental and modeling results. *Metals and Materials International*, 2016, 22: pp. 817–824.

38. Vaßen, R., et al., Vacuum plasma spraying of functionally graded tungsten/EUROFER97 coatings for fusion applications. *Fusion Engineering and Design*, 2018, 133: pp. 148–156.

39. Kuo, T., Chin, W., Chien, C., Hsieh, Y., Mechanical and biological properties of graded porous tantalum coatings deposited on titanium alloy implants by vacuum plasma spraying. *Surface and Coatings Technology*, 2019, 372: pp. 399–409.

40. Movchana, M., Rudoy, Y., Composition, structure and properties of gradient thermal barrier coatings (TBCs) produced by electron beam physical vapor deposition (EB-PVD). *Materials and Design*, 1998, 19: pp. 253–258.

41. Meng, B., Sun, Y., He, X.D., Peng, J.H., Fabrication and characterization of Ni-YSZ anode functional coatings by electron beam physical vapor deposition. *Thin Solid Films*, 2009, 517: pp. 4975–4978.

42. Bhushan, B., *Encyclopedia of Nanotechnology*, 2012, Springer, Netherlands.

43. Tanaskovic, D., Jokic, B., Socol, G., Popescu, A., Mihailescu, I.N., Petrovic, R., Janackovic, D.J., Synthesis of functionally graded bioactive glass-apatite multistructures on Ti substrates by pulsed laser deposition. *Applied Surface Science*, 2007, 254: pp. 1279–1282.

44. Deng, C., Kim, H., Ki, H., Fabrication of functionally-graded yttria-stabilized zirconia coatings by 355 nm picosecond dual-beam pulsed laser deposition. *Composites Part B*, 2019, 160: pp. 498–504.

45. Basu, S.N., Kulkarni, T., Wang, H.Z., Sarin, V.K., Functionally graded chemical vapor deposited mullite environmental barrier coatings for Si-based ceramics. *Journal of the European Ceramic Society*, 2008, 28: pp. 437–445.

46. Abedi, M., Abdollah-zadeh, A., Vicenzo, A., Bestetti, M., Movassagh-Alanagh, F., Damerchi, E., A comparative study of the mechanical and tribological properties of PECVD single layer and compositionally graded TiSiCN coatings. *Ceramics International*, 2019, 45: pp. 21200–21207.

47. Damerchi, E., Abdollah-zadeh, A., Poursalehi, R., Mehr, M.S., Effects of functionally graded TiN layer and deposition temperature on the structure and surface properties of TiCN coating deposited on plasma nitrided H13 steel by PACVD method. *Journal of Alloys and Compounds*, 2019, 772: pp. 612–624.

48. Sato, M., Tu, R., Goto, T., Ueda, K., Narushima, T., Preparation of functionally graded bio-ceramic film by MOCVD. *Materials Science Forum*, 2010, 31–32: pp. 193–198.

49. Shiraishi, N., et al., Effect of functionally-graded calcium titanate film, prepared by metal-organic chemical vapor deposition, on titanium implant. *Applied Sciences*, 2019, 9: pp. 172.

50. Kang, Y.S., Sharma, S.K., Sanders, J.H., Voevodin, A.A., Finite element analysis of multilayered and functionally gradient tribological coatings with measured material properties. *Tribology Transactions*, 2008, 51: pp. 817–828.

51. Joshi, S.C., Ng, H.W., Optimizing functionally graded nickel-zirconia coating profiles for thermal stress relaxation. *Simulation Modelling Practice and Theory*, 2011, 19: pp. 58–598.

52. Reinoso, J., Paggi, M., Rolfes, R., A computational framework for the interplay between delamination and wrinkling in functionally graded thermal barrier coatings. *Computational Materials Science*, 2016, 116: pp. 82–95.

53. Hardwicke, C.U., Lau, Y., Advances in thermal spray coatings for gas turbines and energy generation: A review. *Journal of Thermal Spray Technology*, 2013, 22: pp. 564–576.

54. Reddy, N.C., Kumar, B.S.A., Reddappa, H.N., Ramesh, M.R., Koppad, P.G., Kord, S., HVOF sprayed Ni_3Ti and $Ni_3Ti+(Cr_3C_2+20NiCr)$ coatings: Microstructure, microhardness and oxidation behaviour. *Journal of Alloys and Compounds*, 2018, 736: pp. 236–245.

55. Reddy, N.C., Koppad, P.G., Reddappa, H.N., Ramesh, M.R., Babu, E.R., Varol, T., Hot corrosion behaviour of HVOF sprayed Ni_3Ti and $Ni_3Ti + (Cr_3C_2 + 20NiCr)$ coatings in presence of Na2SO4-40%V2O5 at 650°C. *Surface Topography: Metrology and Properties*, 2019, 7: pp. 025019.

56. Stathopoulos, V., et al., Design of functionally graded multilayer thermal barrier coatings for gas turbine application. *Surface and Coatings Technology*, 2016, 295: pp. 20–28.

57. Lashmi, P.G., Ananthapadmanabhan, P.V., Unnikrishnan, G., Aruna, S.T., Present status and future prospects of plasma sprayed multilayered thermal barrier coating systems. *Journal of the European Ceramic Society* 2020, 40: pp. 2731–2745.

58. Gluhmann, J., et al., Functionally graded WC-Ti(C,N)-(Ta,Nb)C-Co hardmetals: Metallurgy and performance. *International Journal of Refractory Metals and Hard Materials*, 2013, 36: pp. 38–45.

59. Garcia, J., Pitonak, R., The role of cemented carbide functionally graded outer-layers on the wear performance of coated cutting tools. *International Journal of Refractory Metals and Hard Materials*, 2013, 36: pp. 52–59.

5 Fabrication of FGMs by Additive Manufacturing Techniques

*Agnivesh Kumar Sinha, Rityuj Singh Parihar,
Raj Kumar Sahu, and Srinivasu Gangi Setti*

CONTENTS

5.1 INTRODUCTION

Additive manufacturing (AM) implies a process of manufacturing in which material is added in a controlled manner so as to achieve the desired shape and size of the product. AM follows a bottom-up approach unlike subtractive manufacturing techniques which follow a top-down approach in which material is removed in order to develop a product. Fundamentally, AM and subtractive manufacturing techniques are opposite and different from each other. But both techniques are often used simultaneously. Computer numerical control machining, water jet cutting, electrical discharge machining, and laser cutting are a few examples of subtractive manufacturing techniques.

DOI: 10.1201/9781003097976-5

AM involves building products (parts) by the addition of successive layers of materials one by one until the desired dimensions of the product are achieved. Generally, 3D computer-aided design (CAD) models are developed to create parts through various AM techniques. Therefore, AM technique is also termed 3D printing. In its early days, AM was mostly used for the purpose of rapid prototypes which gave it another name, "rapid prototyping". Previously, it was only used to reduce the cost of testing and analysis (by reducing materials, time, and resources) of intricate products by using prototypes developed through 3D printing due to the limitations of AM like type of materials used, infrastructure for large-sized products (design constraints), and wastage due to interruption in process.

From the last decade, AM has been a popular choice of manufacturing technique among materialists and researchers due to its advantages over other conventional manufacturing techniques. AM techniques, unlike conventional subtractive techniques, can handle the development of intricately shaped products in shorter time, being more cost-efficient. Moreover, it also eradicates the use of custom jigs and fixtures which are often used in conventional manufacturing techniques.

AM techniques are now also used in the development of metallic components or products besides plastics. Composites along with functionally graded materials (FGMs) are also being fabricated using various AM techniques. AM in the fabrication of FGMs has made it possible to provide desired material gradations more accurately when compared to conventional manufacturing techniques such as spark plasma sintering which is often used for the fabrication of FGMs. AM is also capable of fabricating symmetrical as well as complex unsymmetrically shaped FGMs unlike spark plasma sintering which is generally used for the fabrication of symmetrical shaped FGMs. These advantages of AM over conventional manufacturing techniques would be beneficial for the exploration and development of various FGMs with different compositions and gradations at a lesser cost in a shorter time with lower wastage of materials resources.

Therefore, this chapter deals with AM and its classifications from the perspective of FGMs where the design and modeling of AM for FGMs are discussed comprehensively. State-of-art material systems used in AM of FGMs are also elucidated. Moreover, the challenges and future scope of AM in the fabrication of FGMs are also briefly discussed which would be beneficial for paving a path for the upcoming researchers and scientists for advancement in this pertinent field.

5.2   DESIGN AND MODELING FOR AM OF FGMs

FGMs include dispersed material properties having gradual variation in chemical compositions, microstructures, and geometry. The AM process methodology involves geometrical modeling, slicing of model, transition in standard tessellation language (STL) form, support creation, and manufacturing. Numerous commercial software packages are available for the 3D printing of multi-material FGMs, such as Autodesk Monolith, voxel-based systems, and Grab CAD. These packages are not capable of fulfilling the industrial demand for desired graded properties due to the existing challenges such as inadequate understanding of multi-material systems, absence of reliable technique, inability to estimate final material properties, and higher costs. The FGM design is a crucial parameter for successful materials preparation, wherein the assignment and description of material characteristics and the nature of voxel (a small unit of 3D volume) are the most significant parameters. This section describes design principles with a focus on geometrical, material, and microstructure attributes.

Geometrical attributes representation is a primary step to the physical visualization of FGMs by lattice design. Mainly four geometrical representation approaches are available in the CAD package, which includes function representation, spatial decomposition, constructive solid geometry, and B-rep [1]. However, in F-rep and B-rep, essential information about material composition and internal structure cannot be defined accurately. Thus modern and flexible geometry representation approaches such as topology optimization, reverse imaging modeling, and voxel basis methods have emerged. Topology optimization (TO) is enforced with mathematical algorithms (solid isotropic materials, homogenization, bidirectional evolutionary structural optimization, and level-set techniques) to achieve the optimum composition gradient using constraints and boundary conditions to get the desired FGMs. A voxel (volumetric pixel) is the smallest elementary unit of volume identical to a pixel in a 2D image. The voxel-design approach is very effective for designing geometrical coordinates and material compositions separately for tailoring material properties. Wherein the geometrical data interpreted in the raster form and used by voxel printer for development of FGMs. In reverse image modeling, CT or MRI data are interpreted in the form of 3D structure and further utilized in the creation of a 3D voxel density pattern by means of algorithms (such as a filtered back projection algorithm). The use of MRI and CT technologies has effectively reduced the time as well as cost incurred in the design and manufacturing of FGMS using AM [2]. In general it involves four steps: image acquisition, data post-processing, virtual constructs design by CAD, and fabrication.

Materials attributes are another important factor in the design of FGMs for AM processes. As FGMs contain multiple materials, thus composition changes gradually and there is an absence of separate boundaries, which restricts the tendency of cracking and/or delamination. In general, designers are focused on the geometrical features of complex shaped products and skip the important material attributes (composition distribution, compatibility, and stability). Hence a fabrication information modeling (FIM) technique was applied to present the geometrical and material attributes together over desired geometry. One of the CAD packages, 3D Euclidean space *E3*, is capable of geometry modeling with homogeneous material. In this software system heterogeneous composition distribution can also be incorporated by using a unit vector. Numerous research works are available to define material distribution in FGMs for AM. Researchers have stored material information in the form of multi-material trees and information regarding heterogeneous material systems extracted directly for fabrication via AM [3]. Also, material distribution is presented in the form of evaluated as well as unevaluated models. Evaluated models represented distribution in discrete and inexact form using voxel models, whereas unevaluated models do not depend upon the intensive spatial, subdivision, decomposition, or discretization [4]. Material distribution can also be defined in three ways: by modifying the conventional geometrical approach for FGMs, by considering composition variation independent from geometry, and by using material primitives, such as points, straight lines, splines, and planes [5].

A microstructural attribute can be achieved by experimental examinations or numerical simulations of microstructure evolution. Microstructure meshing can be developed using backscatter electron diffraction data with DREAM3D code, and statistically comparable demonstration of a microstructure volume element can be created. Also by modeling phase diagrams can be predicted and used to estimate microstructure phase gradient. The phase diagram can be estimated using the CALPHAD software package, wherein thermodynamic computations were assigned to the gradient path model and forecast microstructure phase distribution.

Computer-aided engineering (CAE) and simulation play a crucial role in the optimizing and modeling of AM design processes by estimating properties, performance, and desired geometry of the final product. Simulation can be used to accurately predict the effect of AM process variables on the properties of prepared FGMs and the need for experimentation is avoided. Nowadays finite element (FE) supported CAE techniques are often employed for the simulation of thermo-mechanical processing of AM components [6].

5.3 METHODS FOR AM OF FGMs

5.3.1 DIRECTED ENERGY DEPOSITION

As the name suggests directed energy deposition (DED) is a type of AM technique which produces parts (3D products) directly from a three-dimensional CAD model via energy. In this technique energy is utilized to form a pool of melt on the substrate. Electron beams and lasers are the most common forms of energy used for the deposition of melt on the substrate. This AM technique can be used to produce or fabricate FGMs, alloys, composites, etc. Laser is the most often used energy form in the DED technique for the fabrication of FGM parts by researchers.

DED to create or produce metal components directly from a 3D CAD model by injecting metal powder via a focused high-power laser beam into a molten pool is known as the laser engineered net shaped (LENS) method of AM. A high-powered laser beam is utilized to melt the metal powder which is placed co-axially to the directed laser beam. And the laser beam is focused by one or more lens to a spot to help in the layer-by-layer melting process. There are nine major process parameters in LENS which play a vital role in deciding the properties of FGM parts produced which are the wavelength of the laser, laser power density, metal particle properties, powder delivery rate, powder temperature, overlap deposition, layer thickness, and melt pool temperature. A smaller wavelength of laser helps in avoiding reflectivity issues. Layer thickness decides the power density of the laser. As the thickness of layer increases, the laser gets defocused due to inadequate laser power density to melt powder which leads to flaws like porosity. Thus, laser power is directly related to the surface finish of the part. A high powder delivery rate at high temperature results in large-grain micro-structured parts whereas the particle size of the powder plays a key role in deciding the surface finish of the part. Also, a powder temperature colder than required could result in inadequate fusion whereas too high powder temperature forms plasma. Therefore, it is suggested to maintain the optimum temperature of powder. Overlap deposition may lead to porosity due to the trapped air which affects the surface finish of the parts adversely. Also, a higher melt pool temperature can cause defects, thermal distortion, and higher residual stresses which on cooling lead to cracking. And layer thickness also plays a role in deciding the surface finish of the part. Layers that are too thick may lead to a poor surface finish due laser defocus.

Researchers have prepared Ti-6Al-4V/Inconel 718 FGM by the DED technique utilizing a laser beam as the energy source for the deposition of metal [7]. Figure 5.1 presents the schematic diagram of the DED technique used for the fabrication of FGM in the research. FGMs like Ti–Nb [8] and Ti6Al4V-Mo [9, 10] were fabricated also using the DED technique where it was confirmed that this AM technique is capable of maintaining the required homogeneity. Zirconia coating FGM stabilized with yttria was also successfully fabricated using LENS with a fine microstructure which possessed hardness ranging from 1700 HV to 2000 HV [11]. Functionally

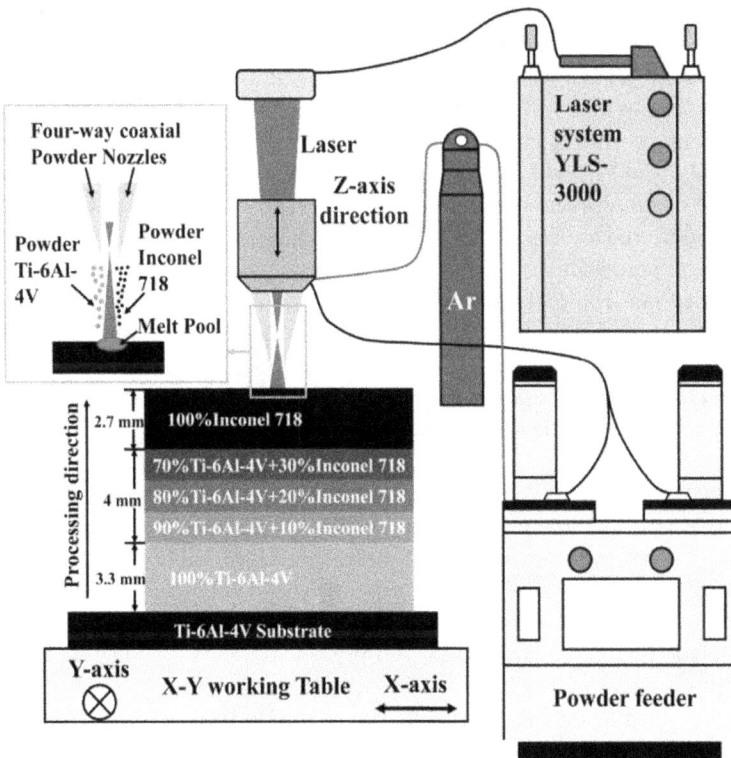

FIGURE 5.1   Schematic diagram of the DED technique used for the fabrication of FGM [7].

graded layers of SS316 L and Ni–Cr–B–Si were developed using 500 W and 1000 W laser power levels so as to deposit the metal powders over the substrate (SS316 L) [12]. An FGM disc of SS316 L/IN625 having a diameter of 80 mm was also fabricated in a study using the LENS technique where laser power was varied from 250 W to 1000 W [13]. However, it was noteworthy that laser power greater than 350 W caused material splashes. 316 L/Inconel718 FGMs with varying gradations were deposited on 304 L in a study using 2000 W laser power to lay a layer of about 0.4 mm [14]. The study manifested that the deposition of a large number of layers caused coarsening of the grains due to large localized heat generation. Similarly, researchers have chosen this AM technique to develop Ti-6Al-4V/TiC FGMs without segregation using separate hoppers where it was found that the flow rate of gas was the least influencing process parameter among four parameters, namely gas flow rate, scanning speed, laser power, and powder flow rate [15]. Researchers also addressed common issues such as deformation and cracking during the fabrication of $Fe_3Al/SS316L$ graded thin wall tubes via the LENS technique [16].

5.3.2 POWDER BED FUSION

5.3.2.1 Selective Laser Melting and Selective Laser Sintering

Selective laser melting (SLM) and selective laser sintering (SLS) are the types of laser power bed fusion AM technique in which powder particles are fused together by laser beam. However, it must be noted that unlike conventional sintering and melting processes all powders are not sintered or melted at the same time in SLS and SLM respectively. SLS generates a layer of predefined geometry of the part by fusing powders with the help of a focused laser beam, whereas SLM refers to the complete melting of metal powders by laser beam. However, it is noteworthy that the process of SLM is referred as SLS in case of polymer materials because completely melted polymers are porous in nature.

Input raw material for the SLS/SLM process is supplied in the form of powder which is fused together using a laser (such as a carbon dioxide laser beam) to create a part of the final product. The complete process of SLM/SLS occurs inside a closed chamber filled with nitrogen gas (N_2) so as to avoid the issues related to the oxidation of powders which might lead to degradation of the material. The process commences with a standard. STL CAD file format which is compatible with most of the 3D CAD model software. This file contains the layered structure of the part. Further, heat-fusible powders in regions corresponding to the 3D CAD geometry model of a single layer are sintered by a moving laser beam to build a part. This focused laser beam is directed towards the powder bed so as to fuse particles of powder together for the selected layer thickness of the part. Layer thickness is approximately 0.1 mm. After the completion of one layer, another layer of loose powders is spread on the surface. In a layer-wise manner, the laser beam fuses the powder particles together to form a solid mass which forms the desired 3D geometry. Un-fused powder particles act as support for the fused region of the part. On the completion of the process, un-fused powders (un-sintered powder) are poured out completely so as to remove the remaining loose powders from the part. During the process, it is required to maintain the temperature of the building platform just lower than the melting temperature (glass transition temperature) of the powder materials for which infrared heaters are often utilized.

The part bed temperature, thickness of layers, laser focus, energy density, and grain shape of the powder are the five important parameters of SLS which play significant roles in deciding the properties of a part. The part bed temperature must be just lower than the melting point of the powder used. The part bed temperature controls the energy required by the laser during the process for the bonding of powder particles. A higher part bed temperature reduces the required incident energy for the fusion of powder during the SLS process. Layer thickness is the depth by which the part piston (building platform) is lowered after the laser scanning of each layer. The incident energy depends on the thickness of the layer to be scanned in the SLS process. A more thick layer will require higher incident energy for fusing the powder particles. However, it is noteworthy that high thickness of layers will generally yield a poor surface finish of the part due to stair stepping (layering defect). Similarly, laser focus is also decided by the thickness of the layer. A longer dwell time of the laser results in a larger melt pool diameter and deeper fusion depth of the layer (thickness of layer). Typically, the layer thickness ranges from 0.1 mm to 0.15 mm. Energy density is the amount of input energy per unit area. It depends on three factors, namely the beam scan speed, laser power, and scan spacing. And all these three factors must be optimized so as to reduce the input energy required to fuse the powder particles together. Powder grain shape largely influences the properties and surface finish of the part during the SLS process. SLS could be used for developing materials like plastics (powdered), metals, and ceramics for their applications in the aerospace, military, electronic, and healthcare industries.

This AM process could be used for the development of intricately shaped 3D parts without trapping the material inside. This is a very fast AM process which is fully self-supporting in nature. In other words, SLS does not require any support for the part during development. The parts fabricated by SLS possess high stiffness and strength. Researchers have successfully fabricated titanium/hydroxyapatite (HA) FGM [17] with the help of SLM having different compositions of HA varying from 0 wt.% to 5 wt.% in FGM with a gradation of 1 wt.% as shown in Figure 5.2. Nano-hardness and micro-hardness values of FGM were found to vary from 5.11 GPa to 8.36 GPa and 3.42 GPa to 5.67 GPa respectively, whereas Young's modulus of Ti/HA FGM increased from 134.24 GPa to 156.26 GPa with an increase in HA content. Also, FGM with a 2% HA ratio showed fracture toughness of 3.41 MPa–m$^{0.5}$. This study showed that SLM could be used for the fabrication of Ti/HA FGM bone implants. However, SLM comes along with its inherent disadvantage of porous surface finish of the fabricated part [18]. Functionally graded Nylon-11/silica nano-composites [19] were also prepared by SLS having 2–10% silica nano-particles. Process parameters were optimized for the fabrication of FGMs. Results showed that with an increase in the volume fraction of silica nano-particles, the modulus increases with decrease in strain, that is, in other words, the addition of nano-particles to FGMs makes them stiffer and brittle in nature. Researchers have also found ways to control the porosity in parts fabricated using SLS [20]. Functionally graded 1D Nylon-11 composites

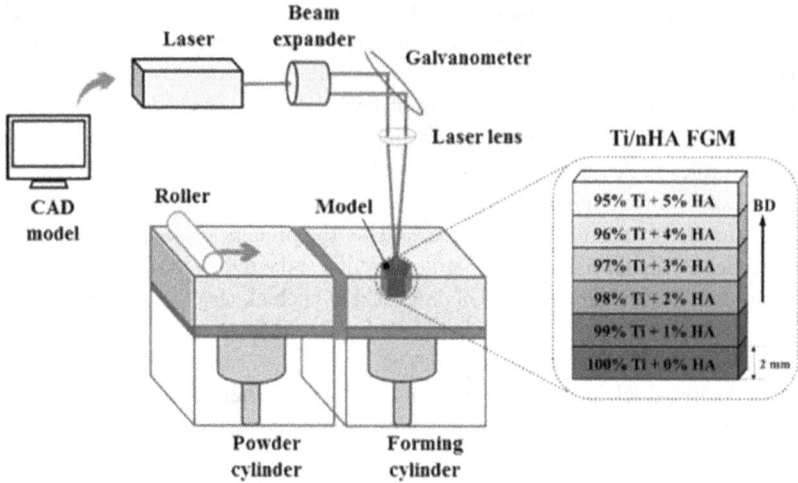

FIGURE 5.2   Schematic diagram of SLM with fabricated FGM [17].

with glass beads were also prepared using SLS in a study which showed that SLS-prepared FGM became stiffer and brittle [21]. Likewise SLM was also employed for the fabrication of W–Cu FGM with a layer thickness of 20 μm, which showed that an inter-diffusion region of about 80 μm was formed [22].

5.3.2.2   Electron Beam Melting (EBM)

EBM involves the melting of material with the help of an electron beam (as the primary source of thermal energy). This AM technique was developed by Acram AB (Swedish-based company) in 1997. The main components of EBM are a grid cup, filament, anode, deflection coil, focus coil, powder container, electron beam, and building table. This process starts with the generation of the electron beam (through filament). The focusing coil present in the EBM setup is utilized to correct the diameter of the beam. The focus coil and deflection coil are unique to EBM. The deflection coil helps in deflecting the electron beam to the powdered material which is to be melted. The fabrication chamber containing the powder container, building table, and focused electron beam is maintained at high temperature and high vacuum. The focused electron beam, targeted by the deflection coil, melts the powder particles present on the building table in small volume within the desired layer thickness which results in the depositing of a layer of metal powder on the building table. Further, the building table is lowered so as to deposit another layer of metal powder on the top of the previously deposited lamina. Finally, numbers of layers are deposited in order to complete the part. And the completed part or component is removed from the fabrication chamber and un-melted excess powder is removed and recycled. Figure 5.3 illustrates a schematic diagram of EBM.

Unlike most of the other AM techniques, EBM printing takes place in a vacuum chamber because some of the metals are highly reactive during the change of their

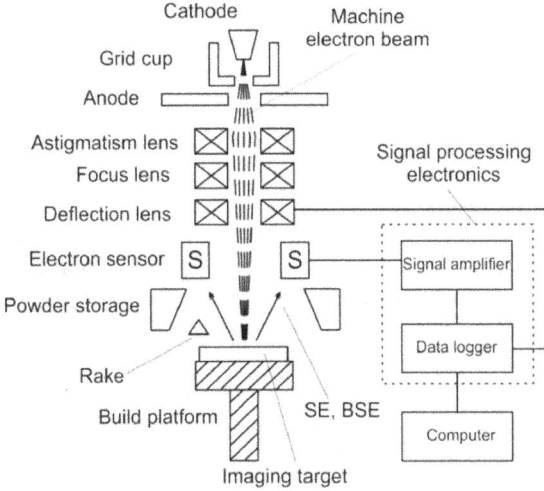

FIGURE 5.3   Schematic diagram of EBM [23].

phases. It also avoids the infiltration of dust particles and atmospheric moisture. EBM is generally used for the fabrication of metallic components as the operating temperature range is very high. The temperature in the chamber itself rises higher than 1000°C. But, due to the use of a vacuum chamber in this process, there is no energy loss or in other words the vacuum helps in maintaining the high efficiency of the process that is almost all of the energy of the generated electron beam is transferred to the material. Energy transfer is five to ten times higher in EBM than the other laser-based AM techniques. EBM is used for the fabrication of biomedical (FGMs based on Ti) and aerospace components.

The properties of parts built using the EBM technique depend on six process parameters, namely beam-focusing time, power density, scanning speed, powder layer deposition, powder properties, and melting temperature. Beam-focusing time is the crucial factor for the change of phase of the powder. If the beam-focusing time is lower than the stipulated time then powder will not be melted properly, whereas, a higher beam-focusing time will make the process slower. Power density determines the amount of energy which will be delivered by the electron beam to the powder. Thus high power density might affect the sub-layer and melt it partially whereas lower power density than the optimal value would cause improper melting which would lead to porosity. Scanning speed is controlled by deflection coils. Higher scanning time leads to improper phase change and lower scanning time might cause defects in sub-layers due to high melting. Powder layer deposition is the thickness of the layer deposited by the electron beam. Higher thickness may lead to lesser beam penetration and if the thickness is too low then the beam might hit the bottom layers, causing defects. Powder should also have capability to absorb the thermal energy for which it is required to pre-heat the powder up to a pre-defined temperature to ensure fast melting.

However, the pre-defined temperature must be optimized because too high a pre-defined temperature could result in longer time of solidification or cooling (lower rate of cooling) which also increases the printing time of EBM. The melting temperature will decide the time required to melt the material. A higher melting temperature causes internal thermal distortion (improper solidification) whereas lower values of melting temperature might lead to the ineffective transfer of heat from beam to powder (improper melting).

In the EBM process, component size or size of the part produced is restricted based on the specification of the deflection coil. It must be noted that EBM uses an optical scanning system which is faster and effective in phase change due to a lack of mechanical movements. Moreover, the resolution of EBM is also higher than SLS. Thus, more intricately shaped components are fabricated using EBM. But the fabrication of FGMs with EBM is much more costly than SLS. Due to the absorption of negatively charged electrons, powder becomes negatively charged which may result in powder particle expulsion from the powder-bed, forming a cloud of powder which is not desirable. Moreover, there might be a chance of repelling the incoming electron beam by the negatively charged powder particles, resulting in a more diffused beam. Thus, the powder particles must be conductive enough to avoid becoming negatively charged. Also, scan strategies should be employed to reduce or eliminate the charging of particles.

Researchers have found that the use of EBM has resulted in reduction of wastage by the recycling of powder [24]. Moreover, a heat source micro-scale model was proposed by researchers [25] for the EBM technique based on the properties of the material, electron beam, and its incident angle. A meso-scale model was also suggested in this research in order to predict the melting process of FGM powder particles during EBM. Simulation of EBM showed no sign of powder scattering phenomenon at 120 W power. Further, a macro-scale fabrication model was developed with the help of effective thermal conductivity which was derived from the meso-scale model. Researchers [26] have also used a new electron beam power bed fusion (EBPBF) approach known as selective evaporation for the fabrication of FGMs. In this approach, light elements could be selectively evaporated using an electron beam from desired regions of the material so as to fabricate FGM from a single material. In other words, the selective evaporation of light material from a single material could result in the creation of desired gradations in order to fabricate FGM. This approach results in achieving a natural transition of interfaces because material is not formed in a layered fashion. This study dealt with the fabrication of a Ti47Al2Cr2Nb-based FGM containing ($\alpha+\beta$) Ti alloy embedded in ($\alpha_2+\gamma$) TiAl alloy. Results also showed that ($\alpha+\beta$) Ti alloy exhibited tensile strength of 1104 MPa along with 8% elongation. The selective evaporation approach in EBM was also used to fabricate FGM having horizontal gradation [27] which is a cumbersome task. 1D FGM was fabricated from a single material powder (Ti47Al2Cr2Nb) containing ($\alpha_2+\gamma$) TiAl alloy and ($\alpha+\beta$) Ti alloy. This study showed that the ultimate tensile strength of FGM decreased on increasing the overlap scan distance.

5.3.3 MATERIAL EXTRUSION BASED

Material extrusion is an AM technique in which a metal or plastic wire is unwound from a coil so as to deposit layers through an extrusion nozzle to complete a part. This extrusion nozzle is heated to melt the plastic/metal. The nozzle is also equipped with a mechanism through which the flow of melt material can be controlled. The extrusion nozzle is mounted on a stage which is movable. The stage can move both in horizontal as well as vertical directions. This makes the movement of the extrusion nozzle easier and hence, the nozzle can be traversed on the table as desirable (as per the geometry of the part) which helps in the deposition of a thin bead of extruded melt to form each layer of the part. Material which is to be extruded through the nozzle must be in semi-solid condition while coming out from the extrusion nozzle. Further, the melt hardens and bonds to the consecutive layer below. The entire process occurs inside a closed chamber held at a temperature just lower than the melting temperature of the material. Once a layer is laid completely, the printing machine will have to move the part downwards or index upwards in order to lay another layer so as to complete the part layer by layer. This metal extrusion process is also popularly known as fused deposition modeling (FDM) as the extruded materials fuse with the already deposited layer so as to form the part completely. A schematic representation of FDM is illustrated in Figure 5.4.

There are two primary approaches to the extrusion technique. The most often followed one is to utilize temperature to control the state of the material. In this approach, the material is melted in a reservoir such that it must be able to flow out through the nozzle and get fused together with the adjoining layer of material prior to its solidification. And another approach is to use a chemical change to start the solidification process. In this approach, a curing agent, residual solvent, drying of wet material, or reaction with air is employed to fuse or bond the materials together. Therefore, a part must be cured or dried completely so as to attain stability. This approach might be more suitable for the fabrication of biochemical parts where materials are required to exhibit biocompatibility with living tissues/cells and hence, the alternatives for material are less or limited. The liquefier temperature, stand-off

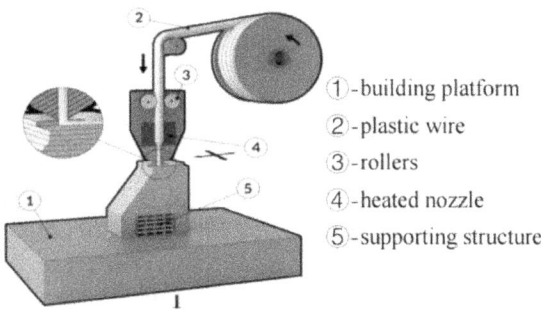

1 -building platform
2 -plastic wire
3 -rollers
4 -heated nozzle
5 -supporting structure

FIGURE 5.4 Schematic representation of FDM [28].

distance, chamber temperature, filament feed rate, deposition speed, nozzle diameter, and type of materials used are the main process parameters in the FDM process. The accuracy of FDM greatly depends on the minimum distance the extrusion nozzle can travel vertically. A dimensional accuracy of about 0.127 mm could be achieved with the help of the FDM process. However, it must be noted that higher accuracy comes at the expense of processing time or printing time. This AM process is usually employed in the aerospace, automotive, manufacturing, medical and health care, architecture, consumer goods, fashion, education, and research industries, etc. FDM is used for the development of prototypes, complex-shaped jigs, and fixtures, intricate molds and tools, etc.

FGM was developed in a study [29] by varying the density of pores in a cellular structure to maintain the desired gradation in FGM. For this purpose, it is necessary to have accurate and precise control over the dimension and geometry of pores in FGM because the size and shape of pores define the density. And the impact strength of a material depends on the relative density. The FDM technique enabled the researchers to have control over the shape and size of pores in the cellular structure so as to develop functionally graded cellular structures. Further, the effects of square, cylindrical, hexagonal, and triangular pore shapes were evaluated according to the energy absorption characteristics and compression strength of FGM. The results showed that hexagonal-shaped pores performed the best among other FGMs with different shaped pores. FGM with hexagonal pores exhibited a compressive strength of 58.2 N/mm^2 and absorbed 27.62 J of energy. This type of functionally graded cellular structure could be used for automobile prototypes, sports, and aero-industries, etc.

Researchers have also employed FDM for functionally grading Al/Al_2O_3 composite patterns in the investment casting process [30]. In another study [31] FDM was used to bond aluminum nitride (AlN) powder with 30 μm particle size and boron nitride powder with 10 μm and 30 μm particle sizes. For this purpose, modified polycaprolactone (PCL) was used as the polymer matrix for bonding the metallic powder together by FDM. Likewise, low-density polyethylene (LDPE)-based FGMs reinforced with SiC and Al_2O_3 particles were also developed [32] where the feed stock filament of FGMs was fabricated by the FDM process. Researchers have also successfully proposed the use of CAD modeling and ANSYS software for the FDM process so as to reduce deformation under transverse loading conditions [33].

5.3.4 STEREOLITHOGRAPHY

Stereolithography (SL) is now popularly referred to as SLA which refers to the stereolithography apparatus. This AM process converts photosensitive liquid into a 3D solid part in a layer-by-layer manner utilizing a high-power laser which convert monomers into cross-linked polymers by means of the polymerization process. In general SLA comprises a tank containing liquid photosensitive monomers, an ultraviolet laser, a movable platform immersed in the tank, and a computer interface which controls the movement of the platform and laser throughout the process of fabrication of the part. Figure 5.5 shows a schematic representation of SLA.

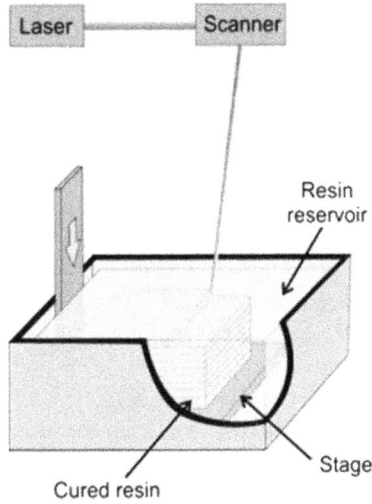

FIGURE 5.5   Schematic representation of SLA [34].

Like every other AM process, the SLA process begins with the generation of a 3D model using CAD software which is the digital representation of the desired part or component. The CAD file is then transformed into a standard tessellation language or standard triangle language (STL) file which is the native and compatible file format for SLA. The laser lays the first layer of the part in a photosensitive polymer resin. The incident laser beam solidifies the liquid resin. The laser is directed and controlled by computer-controlled mirrors. After laying first layer, the platform is raised as per the required thickness of the layer and the additional resin is allowed to flow down below to the printed layer part. Again the laser is used to solidify the upcoming layers of the part to complete the desired part. The fabricated part is cured in an oven for strengthening. The free surface method and fixed surface method are two modes (modalities) of SLA for the purpose of solidification of monomers. As the name suggests, the surface is exposed to air in the free surface method and the solidification occurs at the air/resin interfaces. In this type of mode, precaution measures should be taken in order to minimize waves in the liquid monomer surface as wave formation in the liquid could result in compromising the final dimensional resolution. In this mode, the part is fabricated from the top down. Whereas in the case of fixed surface mode, the resin is stored in a container with a window plate which is transparent for exposure, and solidification takes place at the stable resin/window interfaces. Also, the part is built from the bottom up in the fixed surface mode.

There are two fundamental process variations of SLA, namely, scanning stereolithography and projection stereolithography. In scanning stereolithography, the laser beam moves on the surface and part building occurs in a line-by-line and point-by-point manner with the help of an STL file. However, this variation of SLA may cause a curling effect in parts which results in deviation from the desired dimensions of the part. Thus, to avoid this curling effect in parts, the projection stereolithography variation of SLA came into existence. In projection stereolithography, a digital projector screen is used in place of a laser for flashing a single image of each and every layer of part across the platform such that each layer will be composed of square pixels. This variation is also known as digital light processing (DLP). In DLP, the resolution of the printer corresponds to pixel size whereas in the case of scanning stereolithography, spot size corresponds to the resolution of the printer. SLA is preferred over other AM techniques because it results in a smooth printing surface, high dimensional accuracy of parts, and it is also capable of building high-volume parts of the order of about $50 \times 50 \times 50$ cm^3 without compromising on precision. However, the printing time and cost are higher in SLA. Moreover, steep slopes and overhangs need support in SLA. Unlike other AM techniques, SLA does not offer wide range of material and color alternatives.

There are four vital process parameters in SLA which play vital role in deciding the properties of the printed part which are laser–resin interaction, irradiance and exposure, photo speed, and time scale. Irradiance is the input radiant power of laser per unit area for the conversion of monomers into cross-linked polymers. Exposure is defined as energy per unit area (time of exposure). Laser–resin interaction depends on the line width (thickness of the beam) which is proportional to the spot size of the beam. Cure depth is also controlled by the line width. The depth of the beam is parabolic in nature but for simplicity it is assumed to be a straight line and equal to the thickness of the layer. Photo speed or scanning speed is the scanning velocity which controls the cure depth. The higher the scanning velocity, the greater will be the curing depth due to greater exposure time. The time taken by a laser photon to travel to a photosensitive polymer is in the order of a picosecond (10^{-12} sec). Thus, SLA is a very sensitive process compared to other AM processes as a small deviation in one of these process parameters could result in large deviations in the dimensions of the resulting part due to the liquid state of the polymer.

An Al_2O_3–Si_3N_4 FGM ceramic part/component was fabricated using the stereolithography technique where the [35] Si_3N_4 gradient was set to 20% by volume to achieve six gradations of Si_3N_4 in the Al_2O_3–Si_3N_4 FGM. For the fabrication of FGM, the stereolithography process parameters, namely laser power and scanning speed, were optimized. A study also established the mathematical expression for the viscosity of gradations in FGM in terms of minimum voidage and content of Si_3N_4 in the FGM plate. Prepared gradient combinations showed no sign of delamination or warpage which indicated good bonding between layers. Another study [36] on stereolithography confirmed that SLA is a better AM technique than FDM for the fabrication of FGMs. Due to issues pertinent to filament trajectory, optimal design could not be achieved in the case of FDM. In the SLA technique, CAD models were directly used in topology optimization to achieve optimal design. This approach enables us to tailor FGMs to achieve the desired mechanical strength. However, this approach is not advised for printing resolutions higher than a millimeter scale for printing unit size. Researchers have also used SLA for the fabrication of FGM using the digital light processing (DLP) variation of SLA [37] in one single step for which Fe_3O_4 particles with 300 nm diameter were embedded in FGM mixed with diluted photocurable polymer resin. In this approach magnetic field assistance was needed for which three configurations of magnet were integrated in a magnet holder above the printing bed region. This helped in controlling the distribution of Fe_3O_4 particles in the FGM. The study also showed that the density or distribution of Fe_3O_4 particles depends on magnet strength and the distance between the print part and magnets. However, in the fabrication of FGM by SLA, multi-vat systems were used by researchers which make the process of printing slower due to the inclusion of an additional cleaning step in between the process of changing the material system during part printing.

5.3.5 MATERIAL JETTING

Material jetting is one of the AM techniques which is very similar to 2D inkjet printers in operation. As the name suggests, material is sprayed in droplet form so as to deposit layers one by one to form the complete part in material jetting whereas in inkjet printing, only a single layer of ink droplets is sprayed. In the material jetting process, a print head capable of moving horizontally is used to dispense photosensitive material in the form of droplets which tend to solidify under ultraviolet rays. Likewise, the part is completed in a layer-by-layer fashion. In general, thermoset photopolymers (acrylics) are used as photosensitive material in the material jetting process. Sometimes it is also called the "drop on demand" method which is exclusively used for the fabrication of investment casting patterns.

Initially, liquid resin is heated up to the range of 30°C to 60°C to obtain optimal viscosity for printing. Then, photopolymers are jetted/sprayed in the form of small droplets on the desired region by the print head which travels over the built platform. Further, deposited droplets of photopolymer are cured with the help of an ultraviolet light source which is attached to the print head. Ultraviolet light helps in the solidification of the photosensitive polymer which in turn forms a layer of the part. On completion of a layer, the build platform moves downwards (along the z-axis) by a height equal to the thickness of the layer so as to lay the adjacent layer. In this way the desired number of layers is laid in order to complete the whole part.

Unlike SLA, material jetted parts do not need additional post-curing to achieve optimal properties due to the very small height of parts. Moreover, it also results in a smooth surface of parts with low wastage of materials. It possesses high accuracy of deposition of droplets which majorly depends on the accuracy of the x- and y-axes of the print head. This process also allows multiple materials and colors under one process. Parts printed by material jetting exhibit homogeneous thermal and mechanical properties. However, it is noteworthy that parts produced by material jetting have poor mechanical properties and are brittle in nature. Thus, parts produced by material jetting are not suitable for load-bearing components. And, since parts are made of photosensitive materials, their mechanical properties deteriorate over time. The material jetting process is costly due to its equipment and materials. Materials are limited to polymers, wax, and plastics which is one of the major disadvantages of material jetting. Moreover, it also requires support structures. Parts are also susceptible to warping due to elevated temperature.

There are eight important process parameters, namely diameter of the nozzle, distance between the nozzle and substrate, pressure (to overcome nozzle wall friction), density of material, dynamic viscosity, material velocity, surface tension, and the position accuracy of the x, y, and z motion stage, which govern the properties of the part produced. The material jetting process is generally used for producing parts for dentistry, the medical industry, and jewelry due to the high dimensional accuracy of the process. It also allows the printing of large-volume parts without compromising on dimensional accuracy. It is also one of the fastest AM techniques.

This AM technique also has the capability to fabricate multi-material components (FGMs). Researchers have evaluated the fatigue properties of FGMs fabricated using the material jetting process where it was observed that the material jetted FGM part [38] exhibited poor properties at the interfaces. The study also manifested the influence of material gradient patterns (stepwise or continuous) and the material transition region on the fatigue life of FGMs. Results also showed that the stepwise gradient in FGM components exhibited higher fatigue life and failure through the interface regions. Moreover, it was also observed that FGMs with a shorter material gradient transition length tended to have better fatigue properties than the FGMs with a longer material gradient transition region. Researchers have also successfully fabricated meso-scale (1 mm) digital material using material jetting [39] by regulating the pattern of deposition of polymer at voxel-scale (90 μm). And, with the help of digital material, functionally graded material components or structures could be developed at macro-scale (10 μm). Another study [40] on FGM produced by material jetting analyzed its mechanical behavior which was further validated and verified with the help of finite element analysis and tensile test experimentation carried out for the material jetted FGM specimens. The study claimed that the accuracy of a material jetting printer depends on the size of droplets, structure of the product, size of the product, and shrinkage [41].

5.3.6 HYBRID AM

The challenges of AM techniques for the fabrication of FGMs have led to the evolution of hybrid AM techniques. An objective of the deployment of hybrid AM techniques could be the minimization of print time and cost [42]. In hybrid AM, the combination of several fabrication techniques is used to reduce the number of steps in the process of development of FGMs. This is also beneficial for the customization of properties of FGMs. However, the flow or the sequence of steps must be carefully decided in order to achieve the desired properties of FGMs. Researchers have conveniently used the AM technique along with picoject deposition and micro-extrusion to fabricate FGMs [43] in which an "on the fly" in-situ laser system (laser curing) was employed. Figure 5.6 illustrates the experimental setup for a hybrid AM technique used for the fabrication of FGMs. The study showed that the increase in laser passes tends to transform the microstructure into a densely packed one. The curing of deposited material was done using a laser system which offered fast curing time, less than 1 min, whereas the furnace curing of material took more than 2 hours. Further, sintering parameters were also optimized and were found to be

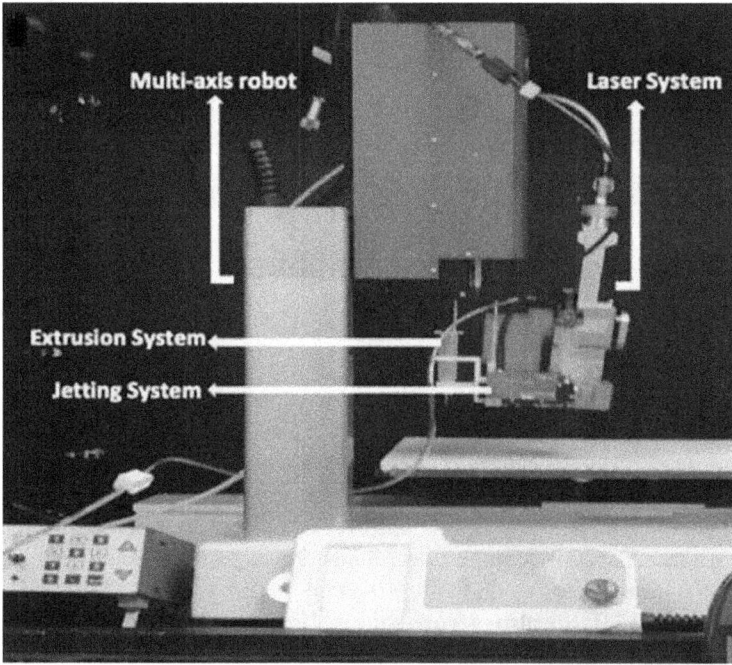

FIGURE 5.6 Experimental setup for hybrid AM [43].

18 W of power with four laser passes. Moreover, FGM was fabricated using the SLS AM technique. It was shown in this research that FGMs could be fabricated using the hybrid AM technique. The use of hybrid AM also made it possible to control the microstructure and hence the properties of FGMs.

Another study [44] on hybrid AM of FGMs involved two techniques, namely cold spraying and SLM. The main objective of the research was to fabricate FGM by the deposition of Al+Al$_2$O$_3$ and Al over Ti6Al4V. Results showed that the Al/Ti6Al4V FGM fabricated using the hybrid AM technique displayed no sign of intermetallic phases at the interfaces (multi-material interfaces). However, it was observed that the Ti6Al4V part fabricated by SLM exhibited defects like large pores whereas the part prepared using the cold spraying technique resulted in the formation of small pores. Figure 5.7 shows the microstructures of the SLM and cold sprayed parts.

5.4 STATE-OF-THE-ART MATERIAL SYSTEMS

AM is a revolutionary manufacturing process and completely changes the scope of FGMs because of the ability to prepare complex shapes with desired composition variation. Therefore, an abundance of research was performed by researchers to fully utilize the AM process for a variety of material combinations according to the area of application. Most of the works available on the Ti alloy, steel, invar FGM used direct energy deposition technique. The titanium alloys have material

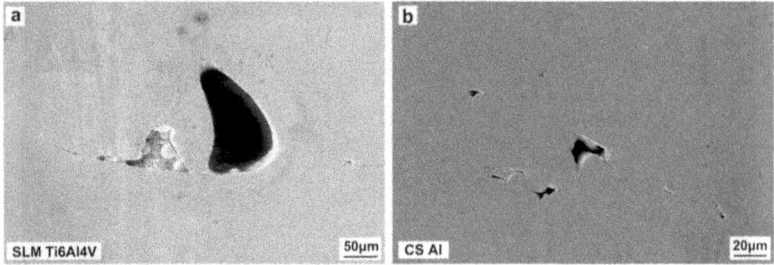

FIGURE 5.7 Microstructure of (a) SLM and (b) cold spraying part with pores [44].

properties such as high strength, superior corrosion resistance, and low density, making titanium alloys the most suitable candidates for applications in aerospace. Thus numerous researchers have tried to prepare FGMs using the AM process. Ti6Al4V/TiC FGMs (with TiC weight fraction variation) were developed by the laser melting deposition process and reported improvements in hardness, tensile strength, and wear resistance. Ti6Al4V/stainless steel joints are very essential in the nuclear and aerospace industries; thus FGMs are prepared by laser-powered AM techniques. Ti6Al4V/invar FGMs are also prepared by laser-powered AM to mitigate the issue of interfacial cracks. Functionally graded Ti6Al4V/Mo without any cracks or observable defects was developed by direct energy deposition. Ti6Al4V/Al$_2$O$_3$ FGMs were fabricated by laser-engineered net shaping to achieve defect-free composition gradients. SS 316L FGM was manufactured by selective laser melting, which contains regions with diverse local functionalities. SS316L/Stellite12 FGMs were prepared by powder injection to achieve high corrosion and wear resistance. In order to use titanium for orthopedic implants it is necessary to reduce stiffness and stress shielding effect. Thus the concept of FGMs was merged with lattice scaffold design, and the designed product was developed by the selective laser melting technique. Invar is an ideal material for processing by AM; therefore it is prepared in the form of an FGM with Invar36/V as well as Inver36/TiC to get a high-strength product. W/Cu is very useful for applications as plasma facing, so it is prepared as FGMs by selective laser melting to reduce interfacial cracks and defects [6].

5.5   CHALLENGES IN AM OF FGMs

AM techniques require the pairing and compatibility of software with hardware in order to work accurately and efficiently. As the first step of AM is the design and modeling of geometry with the help of software, it requires skilled operators or designers so as to understand the given or specified engineering designs of FGMs to create its digital replicas. This hurdle could be overcome by providing suitable training and education to the designers. The types of materials used in each of the AM techniques are different. And all the AM techniques are not compatible with all kinds of materials like metals, polymers, ceramics, and alloys. Thus, one should have a clear understanding of the various types of materials that could be used in specific

AM techniques. This issue limits the applications of AM techniques. Moreover, size constraint is another hurdle for the application of AM techniques in the fabrication of FGMs. Each and every AM technique has its corresponding size constraints for the fabrication of parts.

AM techniques are known for their accurate and precise parts. However, there are some AM techniques which have low resolution and thus result in inaccurate dimensions of built parts in spite of being more expensive than the conventional fabrication techniques. And dimensional accuracy or compositional accuracy plays a key role in deciding the mechanical properties of FGM. An improper gradient might lead to the catastrophic failure of FGMs.

As all the AM techniques produce parts in a layer-by-layer fashion, it introduces anisotropy in the mechanical properties of the parts produced. Hence, two adjacent layers will exhibit anisotropy. Moreover, a slight deviation in the input process parameters of the AM technique could result in a relatively larger deviation in the output geometry and properties of FGMs due to the introduction of heterogeneity (defects). However, surface defects like staircase error might occur due to issues in curve approximation in the STL file. And sometimes STL files are relatively larger than their corresponding CAD files. Moreover, STL files do not have the ability to capture larger curvatures accurately which results in the degeneration of facets.

There are also problems related to gradient distribution and definition of length transition phases in FGMs. It is noteworthy that there is only handful of software packages which are commercially available which have the ability to produce gradients in FGMs (multi-material 3D printing). Moreover, it is also required for the designers to possess complete knowledge of materials behavior so as to fabricate FGMs using AM techniques. Also, it is observed in the literature that the fabrication of FGMs is limited to two-material systems only due to the hardware constraints of AM techniques.

5.6 FUTURE POTENTIAL AND PROSPECTS

There are huge possibilities of advancements in the field of AM of FGMs. But there are some fundamental issues which have resulted in limiting the applications of AM in the fabrication of FGMs. Due to high uncertainty, there are always probabilities of parts being deformed from the desired properties. Thus, it is required to simulate the AM processes to avoid the wastage of resources and to understand the insights pertinent to the uncertainties of the process. Therefore, more simulation tools must be developed.

Another vital hurdle in the fabrication of FGMs using AM techniques is the limited use of materials. For advancement in the field of FGMs, a huge number of compositional combinations of materials must be explored in order to attain the remarkable properties of FGMs. Thus, a materials catalog update would be beneficial. Also, tool paths must be optimized to control the deposition of materials and part geometry which would be beneficial for the minimization of part distortion or deformation. Moreover, real-time process monitoring systems could be employed for the fabrication of FGMs which will further reduce the errors pertinent to processes and hence will result in enhanced productivity.

5.7 CONCLUSIONS

This chapter shows that there are a number of AM techniques available as alternatives for the fabrication of FGMs. Each of the techniques has its own advantages and limitations. However, most of the AM techniques are capable of producing intricate and complex FGM parts. It also shows that the utilization of AM techniques for the fabrication of FGMs is still in progress and has a long way ahead. Also, there is a need to understand the insights pertinent to AM techniques to address the challenges related to the fabrication of FGMs with desired gradations.

This study is an attempt to understand the various types of AM techniques used for the fabrication of FGMs. It also brings out the important process parameters of AM techniques along with their effects on the FGM parts produced. Researchers have also employed unconventional hybrid AM techniques to fabricate FGMs with desired properties (by eradication of errors during fabrication) which has also led to the evolution of AM techniques. However, there is a huge number of barriers apart from those discussed in this study pertinent to material, design, skill, production standard, and non-reliable part production. Various material systems which were developed using different AM techniques were also elucidated. Moreover, it is also observed that most of the AM techniques are flexible and thus preferred over conventional FGM fabrication techniques. Flexibility also helps in achieving the desired intricacy in parts in a precise manner. However, most of the AM techniques are expensive due to their expensive equipment when compared to conventional fabrication techniques. And despite many issues which still remain unaddressed, AM has emerged as an alternative for the fabrication of FGMs which offers limitless potential to be explored.

REFERENCES

1. Ramírez-Gil FJ, Murillo-Cardoso JE, Silva ECN, Montealegre-Rubio W. "Computational Modeling, Optimization and Manufacturing Simulation of Advanced Engineering Materials". In *Advanced Structured Materials*, Muñoz-Rojas P., (eds), Springer, Cham. vol. 49:205–237. 2016. doi:10.1007/978-3-319-04265-7_8.
2. Qian X, Dutta D. Design of heterogeneous turbine blade. *Comput Aided Des* 2003;35:319–29. doi:10.1016/S0010-4485(01)00219-6.
3. Khor KA, Gu YW. Thermal properties of plasma-sprayed functionally graded thermal barrier coatings. *Thin Solid Films* 2000;372:104–13. doi:10.1016/S0040-6090(00)01024-5.
4. Khor KA, Gu YW. Effects of residual stress on the performance of plasma sprayed functionally graded ZrO2/NiCoCrAlY coatings. *Mater Sci Eng A* 2000;277:64–76. doi:10.1016/S0921-5093(99)00565-1.
5. Kim JH, Kim MC, Park CG. Evaluation of functionally graded thermal barrier coatings fabricated by detonation gun spray technique. *Surf Coat Technol* 2003;168:275–80. doi:10.1016/S0257-8972(03)00011-2.
6. Saleh B, Jiang J, Fathi R, Al-hababi T, Xu Q, Wang L, et al. 30 years of functionally graded materials: An overview of manufacturing methods, applications and future challenges. *Compos B Eng* 2020;201:108376. doi:10.1016/j.compositesb.2020.108376.

7. Ji S, Sun Z, Zhang W, Chen X, Xie G, Chang H. Microstructural evolution and high temperature resistance of functionally graded material Ti-6Al-4V/Inconel 718 coated by directed energy deposition-laser. *J Alloys Compd* 2020;848:156255. doi:10.1016/j.jallcom.2020.156255.

8. Schneider-Maunoury C, Weiss L, Perroud O, Joguet D, Boisselier D, Laheurte P. An application of differential injection to fabricate functionally graded Ti-Nb alloys using DED-CLAD ® process. *J Mater Process Technol* 2019;268:171–80. doi:10.1016/j.jmatprotec.2019.01.018.

9. Schneider-Maunoury C, Weiss L, Acquier P, Boisselier D, Laheurte P. Functionally graded Ti6Al4V-Mo alloy manufactured with DED-CLAD ® process. *Addit Manuf* 2017;17:55–66. doi:10.1016/j.addma.2017.07.008.

10. Schneider-Maunoury C, Weiss L, Boisselier D, Laheurte P. Crystallographic analysis of functionally graded titanium-molybdenum alloys with DED-CLAD® process. *Procedia CIRP* 2018;74:180–3. doi:10.1016/j.procir.2018.08.089.

11. Balla VK, Bandyopadhyay PP, Bose S, Bandyopadhyay A. Compositionally graded yttria-stabilized zirconia coating on stainless steel using laser engineered net shaping (LENS™). *Scr Mater* 2007;57:861–4. doi:10.1016/j.scriptamat.2007.06.055.

12. Banait SM, Paul CP, Jinoop AN, Kumar H, Pawade RS, Bindra KS. Experimental investigation on laser directed energy deposition of functionally graded layers of Ni-Cr-B-Si and SS316L. *Opt Laser Technol* 2020;121:105787. doi:10.1016/j.optlastec.2019.105787.

13. Meera Mirzana I, Krishana Mohana Rao G, Ur Raheman S, Zaki Ahmed M. Fabrication and micro-structural study of functionally graded material disc of SS316/IN625. *Mater Today Proc* 2016;3:4236–41. doi:10.1016/j.matpr.2016.11.103.

14. Su Y, Chen B, Tan C, Song X, Feng J. Influence of composition gradient variation on the microstructure and mechanical properties of 316L/Inconel718 functionally graded material fabricated by laser additive manufacturing. *J Mater Process Technol* 2020;283. doi:10.1016/j.jmatprotec.2020.116702.

15. Mahamood RM, Akinlabi ET. Laser metal deposition of functionally graded Ti6Al4V/TiC. *Mater Des* 2015;84:402–10. doi:10.1016/j.matdes.2015.06.135.

16. Durejko T, Zietala M, Polkowski W, Czujko T. Thin wall tubes with Fe3Al/SS316L graded structure obtained by using laser engineered net shaping technology. *Mater Des* 2014;63:766–74. doi:10.1016/j.matdes.2014.07.011.

17. Han C, Li Y, Wang Q, Cai D, Wei Q, Yang L, et al. Titanium/hydroxyapatite (Ti/HA) gradient materials with quasi-continuous ratios fabricated by SLM: Material interface and fracture toughness. *Mater Des* 2018;141:256–66. doi:10.1016/j.matdes.2017.12.037.

18. Kim WR, Bang GB, Kwon O, Jung KH, Park HK, Kim GH, et al. Fabrication of porous pure titanium via selective laser melting under low-energy-density process conditions. *Mater Des* 2020;195:109035. doi:10.1016/j.matdes.2020.109035.

19. Chung H, Das S. Functionally graded Nylon-11/silica nanocomposites produced by selective laser sintering. *Mater Sci Eng A* 2008;487:251–7. doi:10.1016/j.msea.2007.10.082.

20. Sudarmadji N, Tan JY, Leong KF, Chua CK, Loh YT. Investigation of the mechanical properties and porosity relationships in selective laser-sintered polyhedral for functionally graded scaffolds. *Acta Biomater* 2011;7:530–7. doi:10.1016/j.actbio.2010.09.024.

21. Chung H, Das S. Processing and properties of glass bead particulate-filled functionally graded Nylon-11 composites produced by selective laser sintering. *Mater Sci Eng A* 2006;437:226–34. doi:10.1016/j.msea.2006.07.112.

22. Tan C, Zhou K, Kuang T. Selective laser melting of tungsten-copper functionally graded material. *Mater Lett* 2019;237:328–31. doi:10.1016/j.matlet.2018.11.127.

23. Wong H, Garrard R, Black K, Fox P, Sutcliffe C. Material characterisation using electronic imaging for Electron Beam Melting process monitoring. *Manuf Lett* 2020;23:44–8. doi:10.1016/j.mfglet.2019.12.005.

24. Gaytan SM, Murr LE, Medina F, Martinez E, Lopez MI, Wicker RB. Advanced metal powder based manufacturing of complex components by electron beam melting. *Mater Technol* 2009;24:180–90. doi:10.1179/106678509X12475882446133.

25. Yan W, Ge W, Smith J, Lin S, Kafka OL, Lin F, et al. Multi-scale modeling of electron beam melting of functionally graded materials. *Acta Mater* 2016;115:403–12. doi:10.1016/j.actamat.2016.06.022.

26. Zhou J, Li H, Yu Y, Li Y, Qian Y, Firouzian K, et al. Fabrication of functionally graded materials from a single material by selective evaporation in electron beam powder bed fusion. *Mater Sci Eng A* 2020;793:139827. doi:10.1016/j.msea.2020.139827.

27. Zhou J, Li H, Yu Y, Firouzian K, Qian Y, Lin F. Characterization of interfacial transition zone of functionally graded materials with graded composition from a single material in electron beam powder bed fusion. *J Alloys Compd* 2020;832:154774. doi:10.1016/j.jallcom.2020.154774.

28. Zhang C, Chen F, Huang Z, Jia M, Chen G, Ye Y, et al. Additive manufacturing of functionally graded materials : A review. *Mater Sci Eng A* 2019;764:138209. doi:10.1016/j.msea.2019.138209.

29. Duraibabu RV., Prithvirajan R, Sugavaneswaran M, Arumaikkannu G. Compression behavior of Functionally Graded Cellular Materials fabricated with FDM. *Mater Today: Proc* 2020;24:1035–41. doi:10.1016/j.matpr.2020.04.417.

30. Singh S, Singh R. Development of functionally graded material by fused deposition modelling assisted investment casting. *J Manuf Process* 2016;24:38–45. doi:10.1016/j.jmapro.2016.06.002.

31. Wang J, Mubarak S, Dhamodharan D, Divakaran N, Wu L, Zhang X. Fabrication of thermoplastic functionally gradient composite parts with anisotropic thermal conductive properties based on multicomponent fused deposition modeling 3D printing. *Compos Commun* 2020;19:142–6. doi:10.1016/j.coco.2020.03.012.

32. Singh N, Singh R, Ahuja IPS. On development of functionally graded material through fused deposition modelling assisted investment casting from Al2O3/SiC reinforced waste low density polyethylene. *Trans Indian Inst Met* 2018;71:2479–85. doi:10.1007/s12666-018-1378-9.

33. Srivastava M, Maheshwari S, Kundra TK, Rathee S, Yashaswi R, Kumar Sharma S. Virtual design, modelling and analysis of functionally graded materials by fused deposition modeling. *Mater Today: Proc* 2016;3:3660–5. doi:10.1016/j.matpr.2016.11.010.

34. Moritz T, Maleksaeedi S. Additive manufacturing of ceramic components. *Addit Manuf*, Elsevier, 2018:105–61. doi:10.1016/B978-0-12-812155-9.00004-9.

35. Xing H, Zou B, Liu X, Wang X, Huang C, Hu Y. Fabrication strategy of complicated Al2O3-Si3N4 functionally graded materials by stereolithography 3D printing. *J Eur Ceram Soc* 2020;40:5797–809. doi:10.1016/j.jeurceramsoc.2020.05.022.

36. Liu T, Guessasma S, Zhu J, Zhang W, Belhabib S. Functionally graded materials from topology optimisation and stereolithography. *Eur Polym J* 2018;108:199–211. doi:10.1016/j.eurpolymj.2018.08.038.

37. Safaee S, Chen R. Investigation of a magnetic field-assisted digital-light-processing stereolithography for functionally graded materials. *Procedia Manuf* 2019;34:731–7. doi:10.1016/j.promfg.2019.06.229.

38. Kaweesa DV., Meisel NA. Quantifying fatigue property changes in material jetted parts due to functionally graded material interface design. *Addit Manuf* 2018;21:141–9. doi:10.1016/j.addma.2018.03.011.

39. Flores I, Flores I. Design and additive manufacture of functionally graded structures based on digital materials. *Addit Manuf* 2019;30:100839. doi:10.1016/j.addma.2019.100839.

40. Salcedo E, Baek D, Berndt A, Ryu JE. Simulation and validation of three dimension functionally graded materials by material jetting. *Addit Manuf* 2018;22:351–9. doi:10.1016/j.addma.2018.05.027.
41. Yang H, Lim JC, Liu Y, Qi X, Yap YL, Dikshit V, et al. Performance evaluation of ProJet multi-material jetting 3D printer. *Virtual Phys Prototyping* 2017;12:95–103. doi: 10.1080/17452759.2016.1242915.
42. Häfele T, Schneberger J-H, Kaspar J, Vielhaber M, Griebsch J. Hybrid additive manufacturing: Process chain correlations and impacts. *Procedia CIRP* 2019;84:328–34. doi:10.1016/j.procir.2019.04.220.
43. Parupelli SK, Desai S. Hybrid additive manufacturing (3D printing) and characterization of functionally gradient materials via in situ laser curing. *Int J Adv Manuf Technol* 2020;110:543–56. doi:10.1007/s00170-020-05884-9.
44. Yin S, Yan X, Chen C, Jenkins R, Liu M, Lupoi R. Hybrid additive manufacturing of Al-Ti6Al4V functionally graded materials with selective laser melting and cold spraying. *J Mater Process Technol* 2018;255:650–5. doi:10.1016/j.jmatprotec.2018.01.015.

6 Design and Fabrication of a Functionally Graded Model of Bone Using the Fused Filament Fabrication Process

Ankit Nayak, Vivek Kumar Gupta, and Prashant K. Jain

CONTENTS

6.1 INTRODUCTION

The terms rapid prototyping (RP) and additive manufacturing (AM) can be used interchangeably but there is an exception, i.e. RP is used to produce a prototype of a new component whereas AM provides opportunities for both prototypes and final components [1]. AM opens a world of design opportunities for new and complicated

DOI: 10.1201/9781003097976-6

objects which cannot be fabricated using subtractive manufacturing technics, due to their limitations. AM is a technology that produces an object with minimum waste and is very efficient for object-specific low-volume manufacturing. AM is able to process a wide range of materials for part fabrication, and it can be categorized by the form of material used for fabrication. This method can be categorized in three categories: solid-based, liquid-based, and powder-based AM [2].

The AM process can be divided into three parts, i.e. pre-processing (data preparation), processing (part fabrication), and post-processing (part finishing). In AM, the first step is to create the CAD model of the part. The CAD model can be fabricated in various solid modeling packages or by processing point cloud data obtained from various sources like 3D scans or CT scans [3]. This model must be converted into an STL file followed by slicing in multiple layers to generate the path for material deposition. The material deposition rate and the pattern can be controlled in AM. The ability of guided material deposition of AM makes it suitable for functionally graded part fabrication.

Functionally graded material (FGM) has varying properties along its dimensions. Such a component with variation in material density can be fabricated using AM by changing the parameters associated with material deposition and energy density which are required to form a bond in two consecutive layers of material [4]. This property of AM can be used to fabricate objects with heterogeneous build properties and parts with FGM. FGM is suitable for the fabrication of implants. Such implants can be classified as fabricated bone, dental implants, etc. The material properties of such implants are slightly different from the surrounding body parts. The implants in orthopedics are used mostly as structurally enforced artificial bone which is inserted inside the corpus. The current dental implants, composed of a single material, sometimes with a coating layer, are, however, essentially uniform in composition and structure in the longitudinal direction. The conception of FGM may be suitable to apply for an implant.

There are various bones inside the human body which may need to be replaced or repaired at an old age or under certain conditions like injury. But the fabrication of bone with varying porosity and material properties is a complicated task. On the other hand AM is capable of fabricating such a model by supplying the tool path along with varying print parameters. In this chapter, a method for the fabrication of customized parts with heterogeneous material density, using patient-specific CT scan data, is presented.

6.1.1 Additive Manufacturing for Functionally Graded Material

In AM, parts are fabricated by the deposition of material in a layered manner. The material deposition rate and other printing parameters influence the quality and structural behavior of the printed part [4]. The material deposition rate and part printing parameters can easily be changed according to the requirements of the fabricated part. Moreover, the use of functionally graded material in areas like biomedical engineering and tissue engineering is highly appreciated [3]. The fabrication of patient-specific parts with heterogeneous material density is a challenging task. Now it becomes easy with improvements in AM technologies, a wide range of material, and specifically advancements in the nozzle path planning methodology suggested by Jain et al. (2010) [4]. The suggested methodology was able to fabricate parts of varying material properties. Moreover, the presented method is the extended and advanced form of tailoring of material properties suggested by Jain et al. (2010) [4] for the modeling and fabrication of biomedical parts using patient-specific CT scan data. These CT scans are processed to segment out the intended body part in order to make the solid model followed by tool path preparation for AM.

6.1.2 BIOMEDICAL IMAGING

Different techniques like CT scanning and MRI scanning are used for biomedical imaging to analyze the internal anatomy of the human body. CT scanning is based on the principle of X-ray imaging, and the reconstruction of a 3D model from several images (radiographs). A rotary X-ray projector transmit rays throughout the intended body part; these X-rays get attenuated corresponding to the X-ray attenuation coefficient of the body part; when these attenuated rays come in contact with the sensor, it generates an image of the internal anatomy on the basis of the X-ray attenuation coefficient as shown in Figure 6.1 and Figure 6.2.

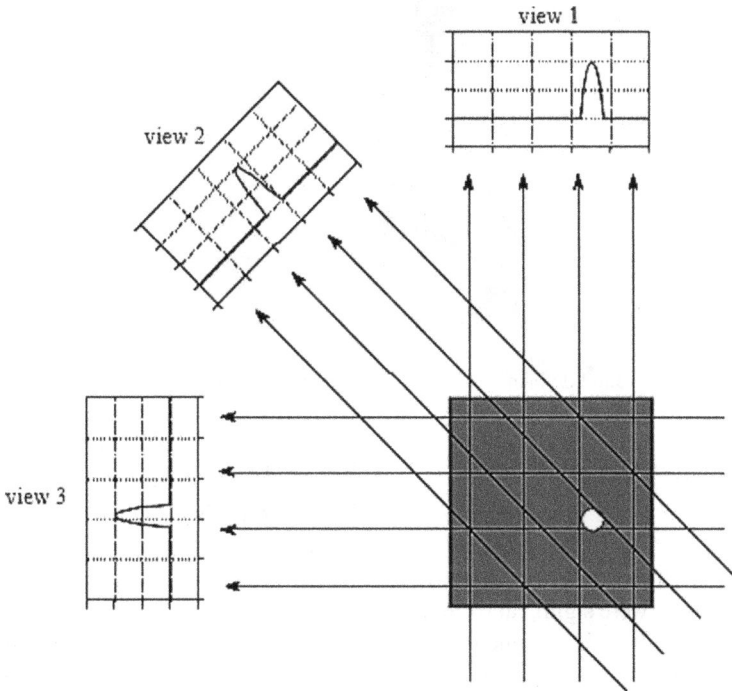

FIGURE 6.1   Formation of an image from attenuated X-rays [5].

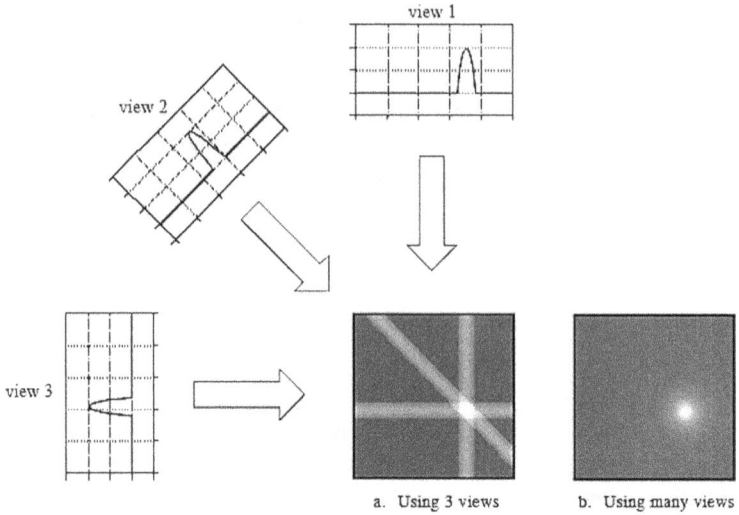

a. Using 3 views   b. Using many views

FIGURE 6.2 Reconstruction from multiple signals [5].

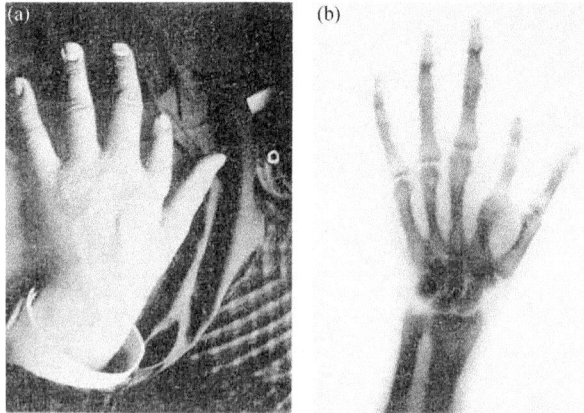

FIGURE 6.3 (a) Hand deformity and (b) same hand seen using radiography.

6.1.3 SEGMENTATION OF THE INTENDED BODY PART

Segmentation of the intended part is the next step towards CAD model preparation of the intended body part. Different organs of the human body have different densities so they attenuate X-rays in relative proportions. It can be observed in Figure 6.3 that the X-rays were significantly attenuated by the bone, hence it is represented with a dark shade, while the attenuation coefficient of the flesh is lower so it disappears in the radiograph. Moreover from Figure 6.3 it can be observed that on the basis of the grayscale value of the pixels, the bone can be segmented out. Different region growing methods are used for the segmentation of the intended body parts. According to the grayscale value which is directly proportional to the X-ray attenuation coefficient the mask for different slices is created. The segmented mask contains the pixels of the specified grayscale value. The masks for an entire stack of the CT scan slices form a point cloud which can be processed to make a solid model of the part.

6.1.4 MEDICAL MODELING FOR ADDITIVE MANUFACTURING

In medical science, information related to the anatomy and the architecture of organs can be gathered using various methods, but the fundamental way is radiography. The varying contrast of a radiograph represents the material density of the corresponding body part (organ). In a digital radiograph, the grayscale value of the pixel is correlated with the Hounsfield value (X-ray attenuation coefficient) of the corresponding body part. On the basis of the grayscale value, the material density of the part can be calculated. The density of the corresponding part will help to prepare the printing parameters and so toolpath. Different slices of a CT scan were analyzed in a customized MATLAB program to calculate the variation in material density of the framer's bone. According to the material density represented in the CT scan, the infill pattern of different layers is decided. To change the part quality and material properties print parameters like nozzle temperature and feed rate are changed. The simultaneous variation in the print density and printing parameters changes the printing density of the fabricated part.

6.2 PROCESS OF FABRICATING FUNCTIONALLY GRADED MATERIAL

The process of part fabrication includes designing the functionally graded part using anatomical data, tool path preparation with varying printing parameters, and fabrication (3D printing). Computer-aided designing of a human body part is called biomedical modeling, and it depends on the anatomical data, acquired from different biomedical image techniques.

6.2.1 BIOMEDICAL DATA ACQUISITION

Biomedical data acquisition is the primary step in designing and fabricating a functionally graded part. Biomedical data involve the architecture and anatomical data of the body part. Anatomical data can be gathered from X-ray imaging methods like CT scans, CBCT scans, or MRI. CT and MRIs are used to capture images within the body at the cross-section. The difference is that MRI uses radio waves and CT uses X-rays to scan the body. The rotating X-ray projector and sensors are used to obtain the complete cross-sectional images in X-ray, which can generate single or multiple layers of slices, whereas MRIs provide extra details on each part with respect to X-ray images; it is used to locate the soft tissues, blood vessels, and bones in various parts of the body. CT scans are frequently used to examine and identify tumors, bone fractures, and cancer development. Whereas MRI scans have been used to identify things such as ankles, brain, joints, wrists, and blood vessels.

CT and MRI imaging uses various formats for transferring files so that there is no loss of data; the images are acquired in a special digital format, the DICOM format (dcm). DICOM format ensures that the high quality of the images is retained. Each CT or MRI scan contains single or multiple images in the DICOM format.

Digital Imaging and Communications in Medicine (DICOM) is a standard format that was created under a non-profit protocol so that it can be used by anyone; it is used for data exchange, the format and structure of images. It was developed by the National Electrical Manufacturers Association (NEMA) for the sharing and visualization of medical images.

DICOM is an international standard for communicating and managing data related to medical images. Its mission is to ensure the safe and secure process of transmission of data. It is used for storing, exchanging, and transmitting medical images all over the globe with its technique of integration of devices for medical imaging from different manufacturers. DICOM images are widely represented as grayscale images and can be represented from pixel value 0 to 255 corresponding to the Hounsfield value of the object which has been scanned under the biomedical imaging system. These images can be processed to acquire data for virtual model reconstruction and additive manufacturing [6, 7].

6.2.2   MEDICAL IMAGE PROCESSING AND DATA EXTRACTION FROM DICOM IMAGES

Medical image processing is a significant technique which has been used for the reconstruction of bone. This application can be used for the solid modeling of body parts, designing of implants, scaffold generation, and demonstration purposes as well. Various techniques have been developed to create artificial bone, directly from STL files or using medical data. In medical diagnostics, computerized tomography (CT) scanning plays an important and essential role in obtaining useful anatomical information. CT scanning gives information about the shape, orientation, and density variation of the bone. Detecting the density variation at the cross-sectional area of the bone is a principal step in modeling heterogeneous parts from CT scan images.

The DICOM file (which is in DCM format) will give the information about the bone density in the form of grayscale. The value of the grayscale will vary from 0 to 255; the darker regions represent less density while the whitest regions represent high density corresponding to the X-ray attenuation coefficient.

A DICOM file contains the sliced image of bone, i.e. the cross-sectional area of femoral bone as shown in Figure 6.4. The white area shows the bone region which has a higher grayscale value as well as high density. The outer region of the bone are muscles and tissues and inner region from the white area is the bone marrow. The white region in the bone is the collection of high pixel values; these pixel values vary from inside to outside which means there is variation in the density of the bone. A different contrast level will help in the segmentation of different parts according to their density and respective Hounsfield values.

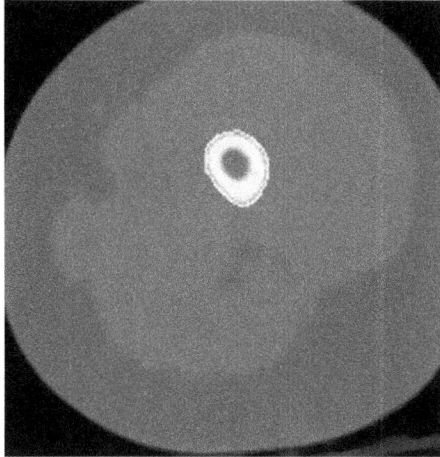

FIGURE 6.4   An image of a random slice of the femur bone.

6.2.3 SEGMENTATION

Two methods have been developed to find the threshold value for segmenting the multiple regions at the boundary point, i.e.

(a) The Analytical Method

The analytical method of segmentation is based on the grayscale value of the boundary pixel having significant contrast value. The maximum difference in neighbor pixels represents the highest contrast value of the image. This value can be used to fix the threshold value for image segmentation by calculating the difference of grayscale value of two consecutive pixels. Several statistical methods or scanning methods can be used to calculate the optimum grayscale value for segmentation. However, in the case of CT scanning it can be calculated on the basis of the Hounsfield value or X-ray attenuation coefficient of the object. The Hounsfield value and corresponding grayscale value of the X-ray image represent the density of the object, and this can be understood from Figure 6.5. The density variation from maximum to minimum is red, blue, green, and yellow respectively. The threshold value of grayscale for the region is 110 to 130 for crayon, 130 to 200 for yellow, 200 to 235 for green, 235 to 238 for blue, and 238 to 255 for red. This variation can be used to segment out the different regions of the density to fabricate a heterogeneous model using AM.

(b) The Graphical Method

The segmented grayscale values obtained on the basis of the graph (Figure 6.6) are 112, 195, 235, 239, and 255.

1. First find out the seed value from which the bone region is separated from muscle and bone marrow. The seed value is obtained using the maximum difference method, and its value is 112.

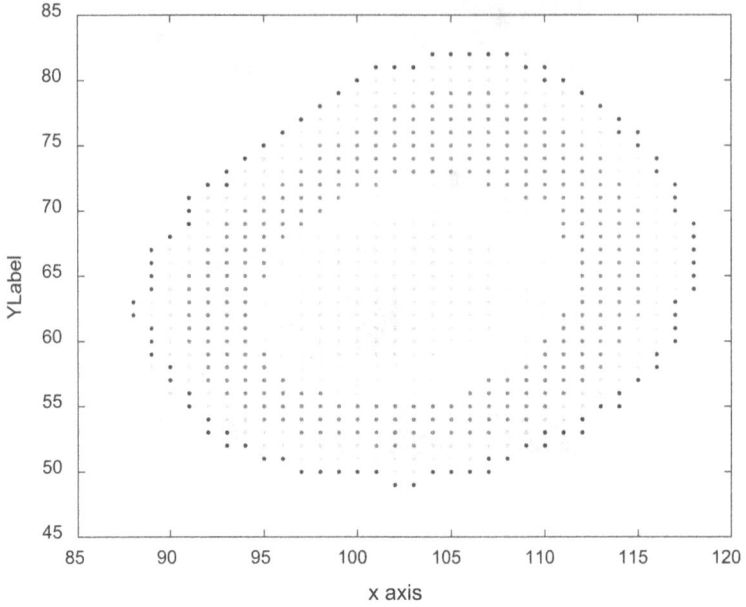

FIGURE 6.5 Segmentation of region using analytical method.

2. Take all the grayscale values of each pixel in this region, arrange them in increasing order, and put them on the y-axis and on the x-axis; the values are arranged according to their order relative to the y-axis.
3. Now the segmented value can be found as:

$$\theta = \tan^{-1}\left(\theta_2\right) - \tan^{-1}\left(\theta_2\right)$$

$$\theta_1 = \frac{B_{n-1} - B_{n-2}}{A_{n-1} - A_{n-2}}$$

$$\theta_2 = \frac{B_n - B_{n-1}}{A_n - A_{n-1}}$$

Where
   θ – difference in angle of two consecutive points
   θ_1 – slope of first line
   θ_2 – slope of second line
   A_1, B_1 – coordinates of first point
   A_2, B_2 – coordinates of second point
   A_3, B_3 – coordinates of third point

The maximum value of θ at each fluctuation of the curve after the seed value is taken as the threshold value for segmentation.

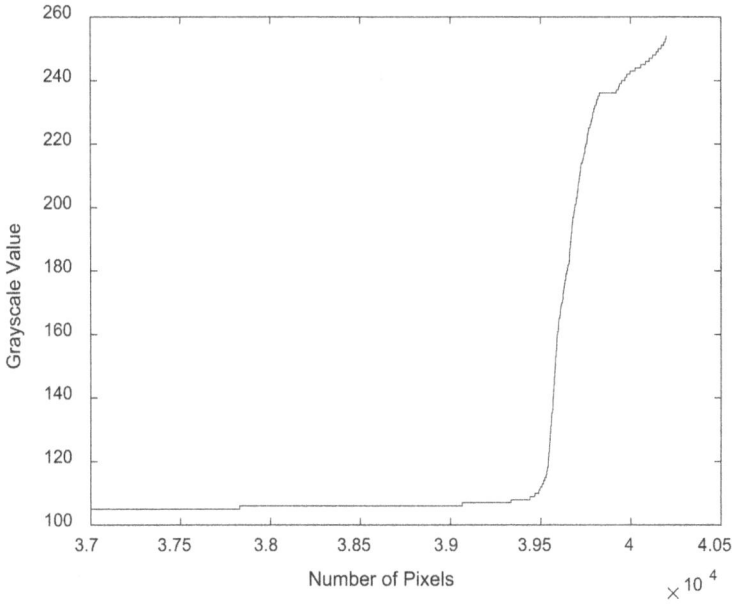

FIGURE 6.6 Graphical analysis.

A DICOM image represents various parts such as bone, muscle, tissue, and bone marrow; however, bone is the region where the largest variety of grayscale is observed due to variation in density. To obtain the region of bone, a segmentation technique is used to get the region of interest. The boundary-value of this region is obtained using the maximum difference method.

The segmentation is performed in three regions, i.e. the blue, green, and red. The red region is denser with a threshold value for boundary separation of 238 to 260, whereas the blue region which is the least dense among these has a threshold value for the boundary of 115 to 200, and the green region has the remaining part. This segmentation value is obtained from the average difference method.

In the outer region, the red and the blue regions are mixed as shown in Figure 6.7. To separate this region a data separation technique is used.

This boundary generation is done by converting all the regions into one and then taking its outer boundary; by removing the segmented region in a sequential manner, the outer shell of the region is obtained.

One slice is taken at a time, applying the segmentation technique based on the threshold value, using the boundary separation technique, and to get a larger localized region there is the further separation of coordinates. The region with high density has a smaller raster gap which means the material will be denser. The raster angle can be 30° and 60° in the anticlockwise direction. After rastering (Figure 6.11), an optimized toolpath is generated. This model shows that,grayscale based boundary separation and raster gap represents heterogeneity in the bone. The density of

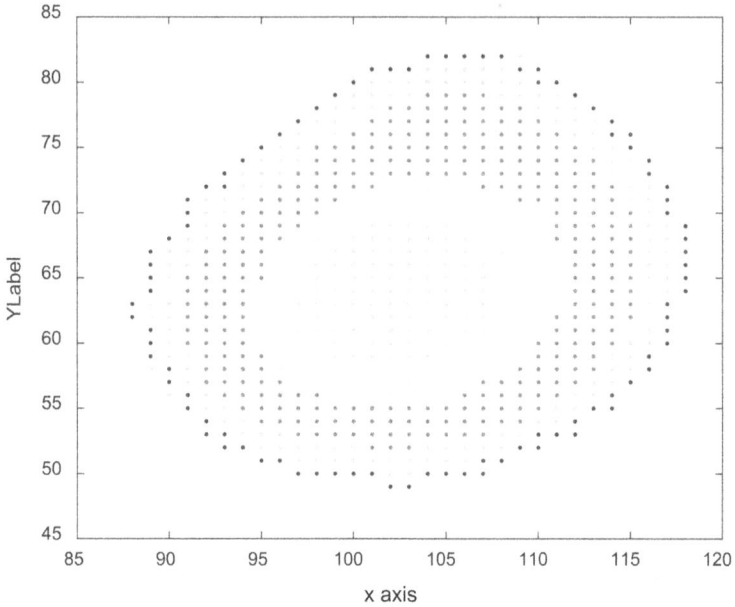

FIGURE 6.7   Segmentation of multi-region.

bone is increasing from the outside to the inside; however, this heterogeneous model explains the relative density with respect to the actual bone.

6.2.4 REGION FORMATION

After image segmentation the bone is divided in different groups according to the range of grayscale values. Whenever there are data values representing the same group or there is a similarity, these values need to be separated from the others. A method has been used which collects the coordinates of those values which can be represented in a certain range. These values store coordinates of each point which comes within a certain range. In Figure 6.7, blue, green, red, yellow, and crayon together represent the bone region. These values are separated from the whole region by using the intensity of each pixel. The range of this segmentation is based on grayscale and the values are 110 to 255.

In Figure 6.7, the number of segmented regions is five, i.e. crayon, yellow, green, blue, and red, whose ranges of values are 110 to 130, 130 to 200, 200 to 235, 235 to 238, and 238 to 255 respectively. Segmented regions such as yellow and green are very small and form a thin line, which will become irrelevant during fabrication due to the lower accuracy of the FFF machine. However, model can be printed using a high-resolution machine according to its applicability, but in this case it has not been considered. To overcome this, the optimization of data is used, and hence the smaller crayon region gets merged with the yellow region, and the crayon region shown here is for the bone marrow which will be used for boundary generation.

6.2.5 CONTOUR FORMATION

A closed contour has been developed from the point set of designated pixel values in order to generate the tool path. In the proposed method the minimum distance of two consecutive points has been measured to form a closed contour. It is the simplest method to obtain a contour, yet it become very difficult when a larger number of points are available and it becomes difficult to choose the next point. A few points may be left behind and a loop will connect in random order to complete the loop. A limiting value of distance should be given to that next point is to include far points in the loop which may be left behind, as stated in previously. The obtained contour will be in the form of a group of connected line segments. The start and end points of the loop were joined to form a closed loop as shown in Figure 6.8.

In Figure 6.8, there is the sequencing of points. These points are randomly distributed. To arrange the points in an order, the nearest neighbor was selected to form a closed loop. Now after finding this line segment, use the head-to-tail method; in this method heads are connected to tails and again the minimum distance method is used to create a closed loop (contour) Similarly, this method has been used for multiple contours as shown in Figure 6.9.

6.3 TOOLPATH FORMATION

Closed contours of different regions represent the boundaries of different segments of the bone with varying density. The density variation can be reflected on

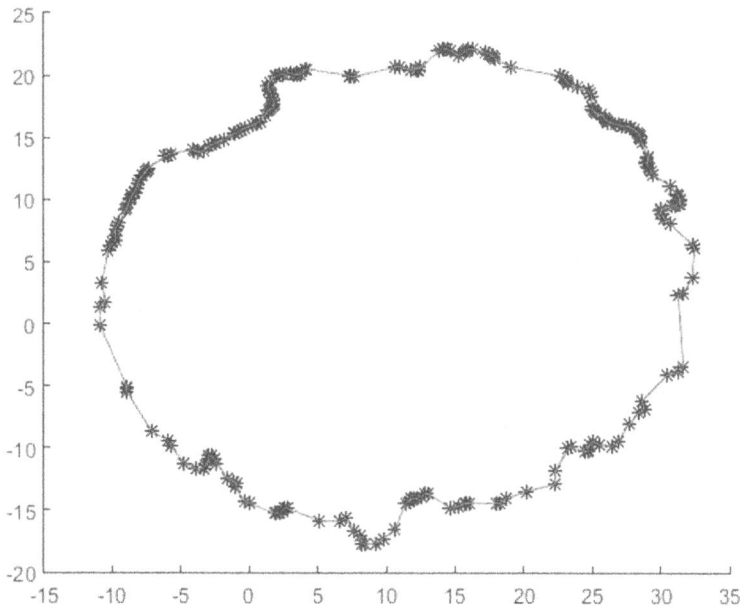

FIGURE 6.8 Sequencing of coordinate points to generate contour.

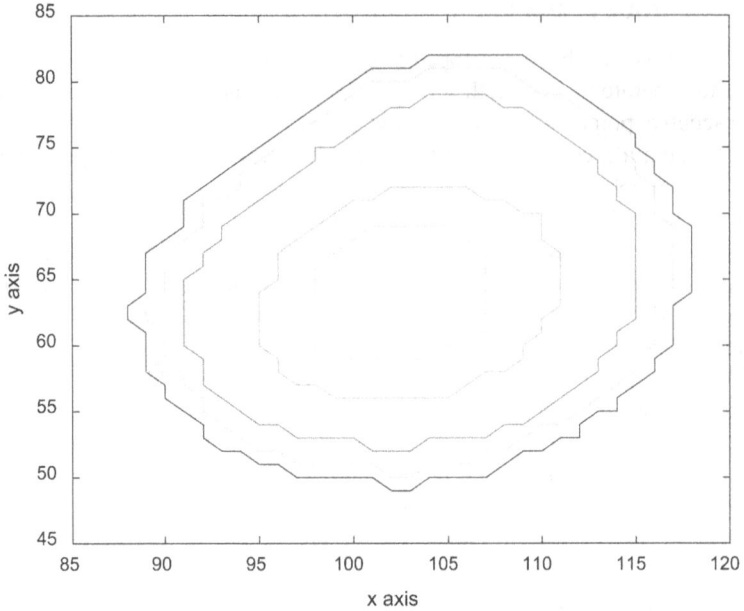

FIGURE 6.9   Contour of multiple regions.

the fabricated part by using different printing parameters like raster gap and raster angle. These variations of the print parameters have been included in the toolpath of the FFF-based AM process.

A standard FDM printing parameter is divided into the few categories. Each parameter has been altered to obtain the optimize design for part fabrication. Here in this model, a single layer of wall boundary is used so that it shouldn't overlap with the infill region. The thickness of each layer is uniform, and each layer has a thickness of 0.4 mm. Each of the DICOM layers which are used here have a gap of 3 mm but the FFF machine, i.e. autoAdobe Duper XL 400, which is used here for part fabrication has a maximum layer thickness of 0.4 mm so for the fabrication process the maximum limiting value of layer thickness is used here. The raster angle used here is 30, 60, 30, and 60 degrees from outside to inside respectively as shown in Figure 6.12. However this raster angle is based on user input, which can be easily changed based on requirements; similarly the raster gap also used, i.e. 0.7, 1.1, 0.5, and 1.5 from outside to inside, can be changed based on the minimum difference in relative density of the model from the original object.

Raster gaps of 0.3, 0.5, 0.3, and 0.7 are designated for a, b, c, and d regions respectively as shown in Figure 6.10. The difference in raster gap shows the variation density in each region.

The toolpath is used to give direction of movement to the extruder head in the cartesian coordinate system. Any toolpath can be generated by connecting all the raster lines which is shown in Figure 6.12. An efficient toolpath should have less extruder movement to reduce the build time. This can be achieved by optimizing the toolpath

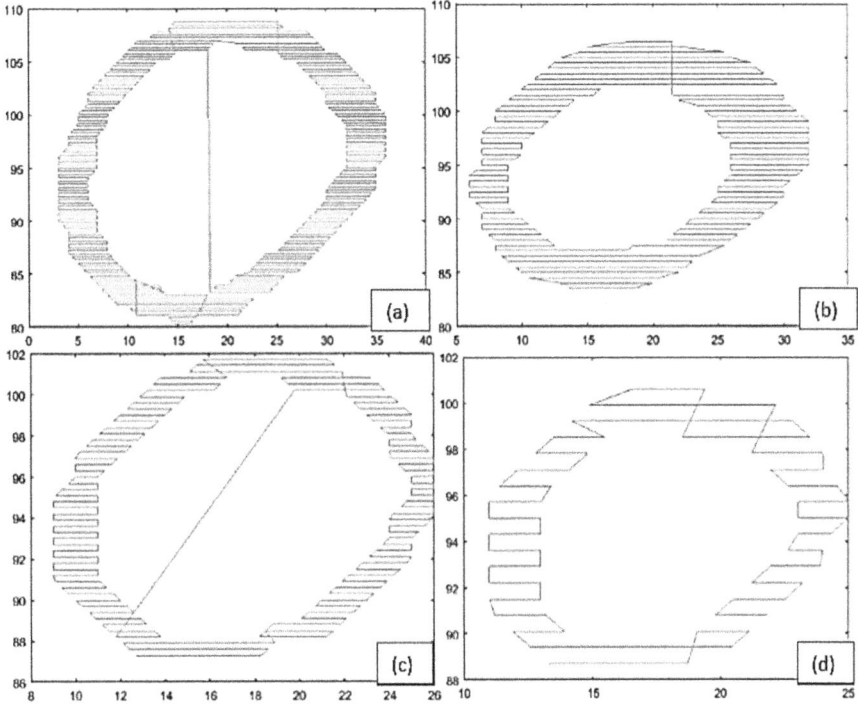

FIGURE 6.10   Separate toolpath of each segmented region.

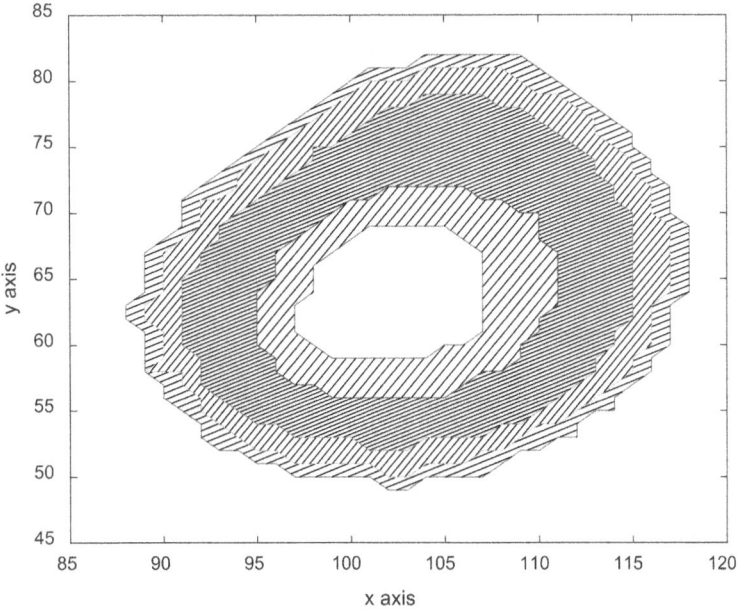

FIGURE 6.11   Rastering using line pattern.

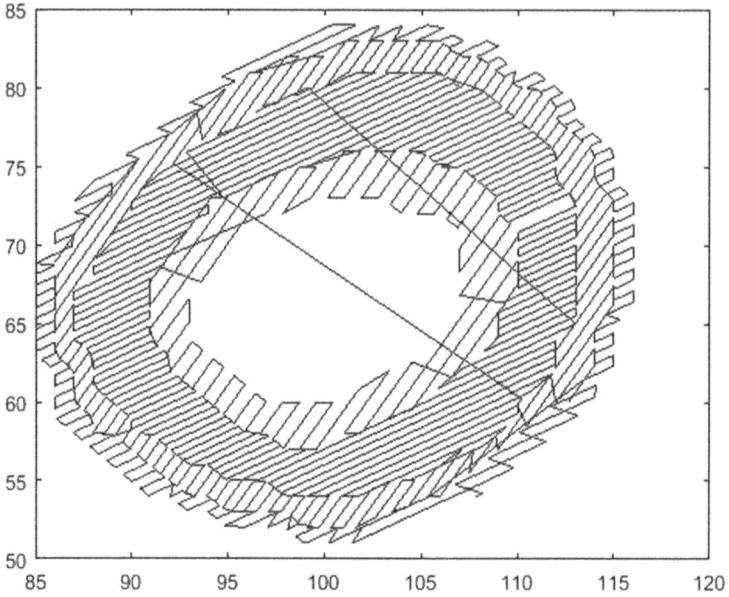

FIGURE 6.12   Toolpath of a single layer.

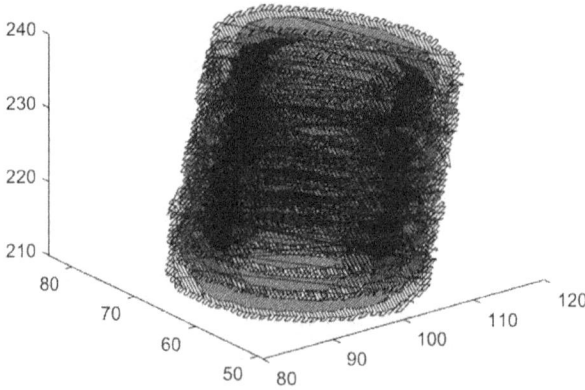

FIGURE 6.13   Toolpath for entire part.

in such a manner that it will move to the nearest raster. By considering this condition a toolpath for the FFF process has been generated as shown in Figure 6.13.

The toolpath has been generated for each layer of the part by following the same algorithm as described earlier. The tool path of different layers was arranged according to the layer height to form a 3D model as shown in Figure 6.13.

This model has been designed to fabricate heterogeneous objects such as bone. There are various bones inside the human body which need to be replaced in old age or under certain conditions. These bones are the hip joint and knee joint. This model has the capability to fabricate not only these bones, but any type of bone in heterogeneous form. Here, in this case a femur bone is used as it has sufficient thickness to observe the heterogeneity within the bone.

6.4 SOFTWARE AND HARDWARE INTEGRATION

An FFF machine has three stepper motors to control the three-dimensional motion of the extruder along the designated path and one motor to control the feed of filament wire to the extruder. These motors are controlled by a dedicated controller which ensures precise motion according to the given commands. In an FFF machine, a nozzle-extruder head is used to deposit material on the build platform. The dedicated controller of the FFF machine is capable of reading the G&M codes. Such codes contain information regarding the motion of different stepper motors. G&M codes are decoded by the controller and pulse trains are supplied to the designated motor according to the G code. The E variable of machine code controls the amount of extrusion according to the layer height and print speed while the X, Y, and Z codes control the position of the extruder. A variable T is designated to control the temperature of the extruder and build platform. These codes include the coordinates of the extruder position and are supplied to the machine for part fabrication.

The code which contains toolpath information plays a major role in the part fabrication. The operating device traces all the coordinates from the toolpath and gives instruction to the nozzle. After that, the nozzle of the FFF machine shown in Figure 6.14 moves through every single point of the toolpath including the idle motions and deposits material on the platform. After the deposition of a single layer,

FIGURE 6.14 FFF machine.

the build platform moves down up to a layer thickness value. The process is repeated until the complete physical model finished.

6.5 APPLICATION

The capability of the proposed approaches has been demonstrated by considering a heterogeneous model for a femur bone used in real-world applications as described below. This model is capable of producing any heterogeneous object whose DICOM image is available.

There are various applications for which the heterogeneous model can be used such as:

1. Hip joint replacement
2. Knee replacement
3. Demonstration purposes

Figure 6.15 shows the fabrication of different numbers of layers for different scale values. This shows that this model can be generalized for single and multiple layers for the fabrication of any heterogeneous objects.

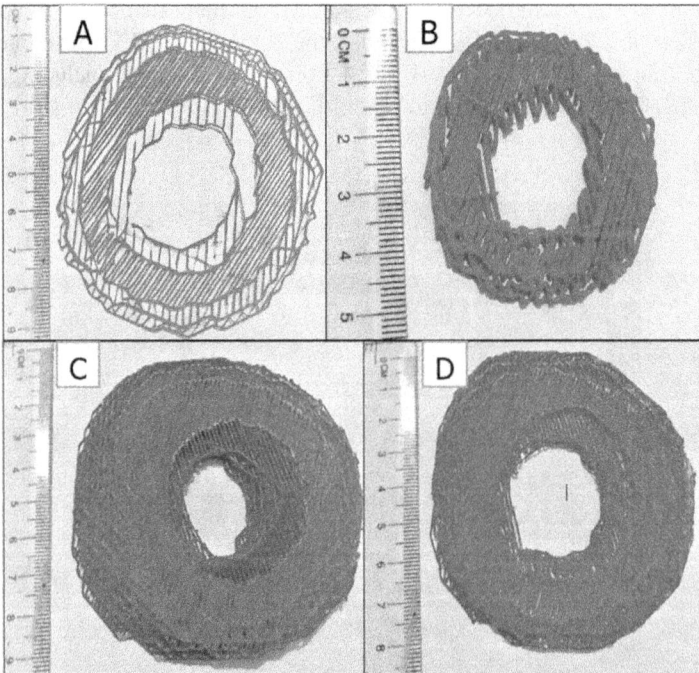

FIGURE 6.15 (A) Single layer at 2.5 scaling, (B) two layers at a scaling value of 2, (C) five layers at 2.5 scaling, (D) three layers at a scaling value of 2.5.

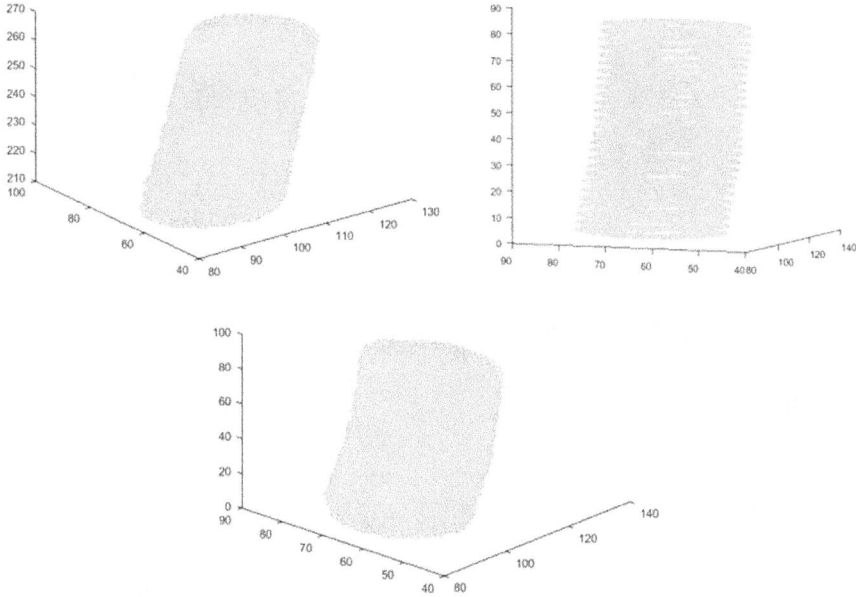

FIGURE 6.16   Different orientation of heterogeneous model.

Figure 6.16 shows the heterogeneous model for multi-slices of a DICOM image. A different orientation of the model is used to compare the similarity between the modeled image and the fabricated image. This fabrication is done on an FFF machine; the machine used is autoAdobe Duper XL 400. Fabricated model shown in Figure 6.17. The working volume of this machine is $200 \times 200 \times 200$ (in mm). The toolpath generated above is used to fabricate this femur bone as shown in Figure 6.16. G and M codes have been implemented on the coordinates of the toolpath.

6.6   CONCLUSIONS

The FFF process is commonly used in the AM process. The software used for density variation through the FFF process and its limitations revealed that localized variation in a part is not possible with available toolpath planning algorithms. Thus, a path planning tool has been developed for providing different set of process parameters in each layer according to the density variation of the part. A detailed explanation of segmentation, infill parameters, and toolpath generation along with its algorithm is presented. Various infill parameters can be implemented on a segmented region to obtain similar variation in density as in the grayscale values of a DICOM image. The feasibility of the proposed path planning tool shows that it can be further improved for advanced material processing AM machines for the fabrication of medical implants.

FIGURE 6.17   Fabrication of femur bone for 20 slices of DICOM.

REFERENCES

1. K. V. Wong and A. Hernandez, "A review of additive manufacturing," *ISRN Mechanical Engineering*, vol. 2012, pp. 1–10, 2012, doi: 10.5402/2012/208760.
2. M. Taufik and P. K. Jain, "Role of build orientation in layered manufacturing: A review," *International Journal of Manufacturing Technology and Management*, vol. 27, no. 1–3, pp. 47–73, 2013, doi: 10.1504/IJMTM.2013.058637.
3. Francis, Vishal, et al. "Influence of 3D printing technology on biomedical applications: A study on surgical planning, procedures, and training," in: *Advances in Materials Processing*. Springer, Singapore, 2020, pp. 269–278.
4. P. K. Jain, P. M. Pandey, and P. V. M. Rao, "Tailoring material properties in layered manufacturing," *Materials and Design*, vol. 31, no. 7, pp. 3490–3498, 2010, doi: 10.1016/j.matdes.2010.02.029.
5. S. W. Smith, *Digital Signal Processing*, vol. 1. Elsevier, 2002.
6. S. Y. Ahn, N. H. Kim, S. Kim, B. Karabucak, and E. Kim, "Computer-aided design/computer-aided manufacturing–guided endodontic surgery: Guided osteotomy and apex localization in a mandibular molar with a thick buccal bone plate," *Journal of Endodontics*, vol. 44, no. 4, pp. 665–670, 2018, doi: 10.1016/j.joen.2017.12.009.
7. A. Nayak, P. K. Jain, P. K. Kankar, and N. Jain, "Computer-aided design-based guided endodontic: A novel approach for root canal access cavity preparation," *Proceedings of the Institution of Mechanical Engineers, Part H: Journal of Engineering in Medicine*, vol. 232, no. 8, pp. 787–795, Aug. 2018, doi: 10.1177/0954411918788104.

7 Recent Advancements in Analysis of FGM Structures and Future Scope

H. D. Chalak and Aman Garg

CONTENTS

7.1 INTRODUCTION

Due to the issues possessed by the classical composite laminates and laminated structures with sandwich configuration such as delamination [1–4], matrix-fiber debonding [5, 6], stress concentration/stress channeling effects [7–9], etc., a new class of materials called functionally graded materials (FGM) was invented by Japanese scientists in 1984. Instead of occurring in the form of layers as in classical laminates, FGMs show a smooth gradation of material property in the desired direction. In the past decade, multi-directional FGM structures have also been developed in which the material property varies regularly in more than one direction [10]. Various structural elements, such as plates, beams, slabs, and shells, constitute a structure. Therefore, the analysis of all these structural elements must be carried out so as to know the complete behavior until failure.

FGMs are manufactured usually using two materials: metals and ceramics. The most widely used metals for FGMs are aluminum, zinc, copper, nickel, titanium, etc. For ceramics, oxides of various metals such as aluminum, zirconium; carbides of various elements like silicon, boron; nitrides such as silicon nitride, etc., are used.

DOI: 10.1201/9781003097976-7

The metal constituent provides toughness and mechanical strength, and the ceramic constituent of the material provides high-temperature resistance due to its low thermal conductivity. Details on the manufacturing of FGMs are available in the work reported by Jha et al. [11]. Among the different processes, powder metallurgy is the most widely used process for manufacturing FGMs.

Various laws are reported by the researchers for defining the properties of FGMs at different points within the structural element. The following are the two widely used schemes available in the literature for determining effective material properties: Voigt's rule of mixture, and the Mori–Tanaka scheme. The power law, exponential law, and sigmoidal laws are the most widely used laws [10–16]. These laws are also modified to develop sandwich FGM structures. A sandwich FGM element is supposed to be made up of three layers in which the top or bottom face is made up of ceramic and the core layer is made up of FGM phase and vice versa.

Homogenization laws for FGMs:

1. Mori–Tanaka scheme: the scheme is mainly used for the discontinuous particulate phase of graded composites. The effective Young's modulus (E), bulk modulus (K), shear modulus (G), and Poisson's ratio (v) can be calculated as:

$$E = \frac{9KG}{3K+G}, v = \frac{3K-2G}{2(3K+G)},$$

$$\frac{K-K_1}{K_2-K_1} = \frac{V_f}{1+(1-V_f)\dfrac{(G_2-G_1)}{G_1+f_1}}, \frac{G-G_1}{G_2-G_1} = \frac{V_f}{1+(1-V_f)\dfrac{(G_2-G_1)}{G_1+f_1}} \tag{7.1}$$

where $f_1 = \dfrac{G_1(9K_1+8G_1)}{6(K_1+2G_1)}$

2. Voigt model: the scheme is based on a simple rule-of-mixture. This scheme is much simpler than the Mori–Tanaka scheme and is the most widely used. The material property of FGM is a function of material property and volume fraction of its constituents, which is defined as:

$$P = \sum_{i=1}^{n} P_i V_{fi} \tag{7.2}$$

where P_i and V_{fi} are the material property and volume fraction respectively of the constituents of FGMs.

$$P = (P_1 - P_2)V_f + P_2 \tag{7.3}$$

Using the above relation (Equation 7.3), the effective properties can be calculated. However, in the literature, it is recommended that for porous

FGM plates, instead of Voigt's scheme, Mori–Tanaka scheme should be used [10, 16].

Chi and Chung [17, 18] carried out a comparative study between the FGM plates made up of three homogenization rules: power-law graded (P-FGM), exponential law graded (E-FGM), and sigmoidal FGM (S-FGM). The following are the three most commonly used gradation laws:

- P-FGM: this law is the most widely used by researchers. The material property variations as per this law for one-directional FGM, two-directional FGM, and sandwich FGM (Equations 7.7 and 7.8) are:

  For 1D and 2D FGMs, the effective material property at any point can be determined as:

$$P_z = \left(P_t - P_b\right)V_f + P_b \tag{7.4}$$

The volume fraction V_f for 1D (Equation 7.5) and 2D FGM (Equation 7.6) can be written as:

$$V_f = \left(\frac{z}{h} + \frac{1}{2}\right)^n \tag{7.5}$$

where n is the power-law exponent

$$V_f = \left(\frac{z}{h} + \frac{1}{2}\right)^{nz}\left(\frac{x}{a}\right)^{nx} \tag{7.6}$$

where nz and nx are the power-law exponents in the z- and x-directions respectively.

Figure 7.1 shows the gradation of material property across the thickness for 1D, 2D, and sandwich FGM structures.

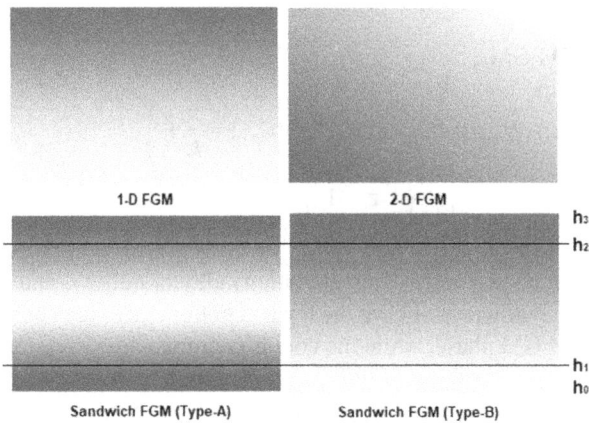

FIGURE 7.1 Material property gradation across the thickness of the FGM structural element.

For sandwich P-FGM structures, Equation 7.4 is used for determining the effective property within the element. Sandwich FGM structures are made up of three layers (Figure 7.1). The volume fraction can be determined as:

- Type A: the sandwich FGM construction is considered to have top and bottom faces of the FGM phase, and the core is ceramic.

$$V_c(z) = \left(\frac{z - h_0}{h_1 - h_0}\right)^n \text{ for } z \in [h_0, h_1], V_c(z) = 1 \text{ for } z \in [h_1, h_2]$$

(7.7)

$$V_c(z) = \left(\frac{z - h_3}{h_2 - h_3}\right)^n \text{ for } z \in [h_2, h_3]$$

- Type B: in this type of construction, the core is made up of FGM phase while the top and bottom faces are ceramic and metal, respectively.

$$V_c(z) = 0 \text{ for } z \in [h_0, h_1],$$

$$V_c(z) = \left(\frac{z - h_1}{h_2 - h_1}\right)^n \text{ for } z \in [h_1, h_2],$$

(7.8)

$$V_c(z) = 1 \text{ for } z \in [h_2, h_3]$$

- E-FGM: this law is used mostly in fracture mechanics [10]. The material property variation as per this law is:
  - 1D FGM:

$$P_z = P_t e^{\left(\frac{1}{h}\right)\ln\left(\frac{P_b}{P_t}\right)\left(z + \frac{h}{2}\right)}$$

(7.9)

- 2D FGM:

$$P_{(x,z)} = P_{(0,0)} e^{k_1\alpha(x) + k_2\beta(z)}, \alpha(x) = \frac{x}{L} + \frac{1}{2},$$

(7.10)

$$\beta(z) = \frac{z}{h} + \frac{1}{2}, k_1, k_2$$

where, $\alpha(x)$ and $\beta(z)$ are gradation index along the x- and z-directions

Sandwich E-FGM: the top and bottom faces are considered to be of FGM phase, while the core is ceramic.

$$V_c(z) = \left(\frac{2z + 1}{2h_1 + 1}\right)^n \text{ for } z \in [h_0, h_1], V_c(z) = 1 \text{ for } z \in [h_1, h_2],$$

(7.11)

$$V_c(z) = \left(\frac{2z - 1}{2h_2 - 1}\right)^n \text{ for } z \in [h_2, h_3], P(z) = P_m e^{\left(\ln\left(\frac{P_c}{P_m}\right)V_c(z)\right)}$$

- S-FGM: this law is the combination of two power laws. The 2D form of this law is not available in the literature.
  1D FGM:

$$f_z = 1 - 0.5 \left[\frac{\frac{h}{2} - z}{\frac{h}{2}} \right]^n \quad \text{for } z \in [0, h/2],$$

(7.12)

$$f_z = 1 - 0.5 \left[\frac{\frac{h}{2} + z}{\frac{h}{2}} \right]^n \quad \text{for } z \in [-h/2, 0]$$

Sandwich S-FGM: the top and bottom faces are made up of FGM and the core of ceramic.

$$V_c(z) = 0.5 \left(\frac{z - h_0}{h_{m1} - h_0} \right)^n \text{ for } z \in [h_0, h_{m1}] \text{ where } h_{m1} = (h_0 + h_1)/2$$

$$V_c(z) = 1 - 0.5 \left(\frac{z - h_1}{h_{m1} - h_1} \right)^n \text{ for } z \in [h_{m1}, h_1]$$

$$V_c(z) = 1 \text{ for } z \in [h_1, h_2]$$

$$V_c(z) = 1 - 0.5 \left(\frac{z - h_2}{h_{m2} - h_2} \right)^n \text{ for } z \in [h_2, h_{m2}] \text{ where}$$

$$h_{m2} = (h_2 + h_3)/2$$

$$V_c(z) = 0.5 \left(\frac{z - h_3}{h_{m2} - h_3} \right)^n \text{ for } z \in [h_{m2}, h_3]$$

(7.13)

Figure 7.2 shows a graphical representation of a volume fraction of ceramic across the thickness of the sandwich FGM structure.

FGMs are most widely used in constructing various structures/structural elements in aerospace and aeronautics; automobiles; defense armaments, e.g. tanks, missiles; structural elements in buildings/bridges; optoelectronic devices; the medical field; sports equipment, etc. (Figure 7.3). Due to the wider applicability of FGMs, it is necessary to study their linear and nonlinear behavior. In special cases such as the defense industry, impact and blast behaviors are of the utmost interest as well as the dynamic behavior. Thus, to accurately predict FGM structures' behavior, several theories are reported in the literature. This chapter summarizes the studies based on the static, vibration, and buckling behavior of FGM structures.

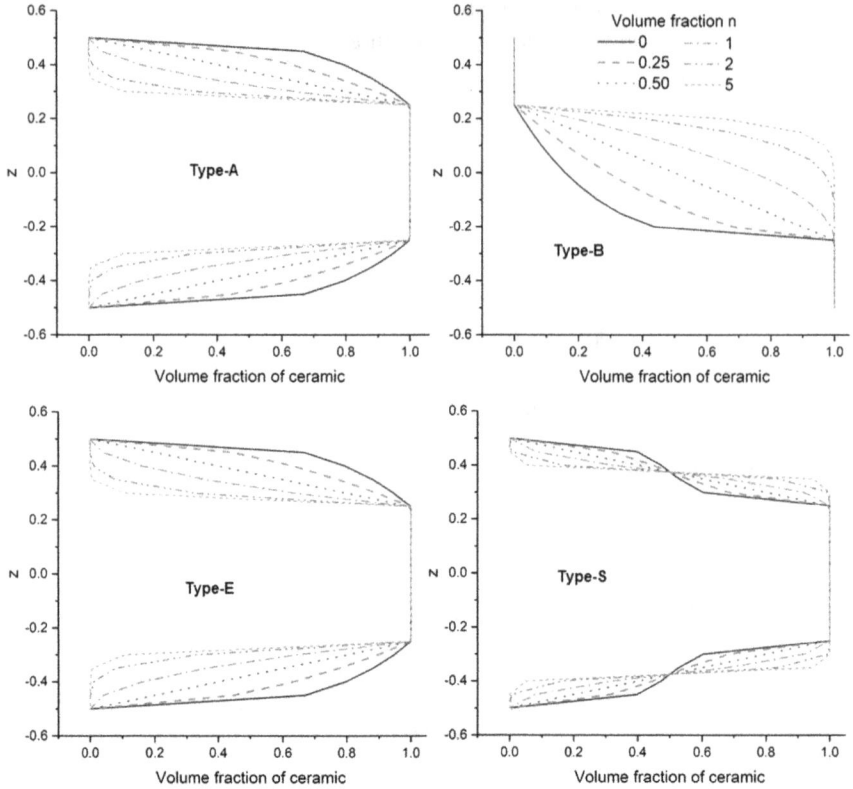

FIGURE 7.2 Variation of the volume fraction of ceramic across the thickness of different types of sandwich FGM structures (1D sandwich FGM).

Various theories are considered for the analysis of FGM structures under various loading cases in the literature. The earliest available theory is called classical laminated theory. However, this theory neglects transverse deformation effects and hence can predict the behavior of thin structures effectively. First-order shear deformation theory (FSDT) assumes a constant transverse displacement field across the structure's thickness. The models based on this theory predict constant transverse shear stresses across the thickness. But actually, the transverse shear stresses are parabolic in nature. To define the transverse stresses accurately, a shear correction factor is required. This shear correction factor depends upon various factors such as end conditions, material properties, thickness scheme, etc. [19–22]. In higher-order shear deformation theories (HSDT), the in-plane displacement field is expanded as a higher-order variation concerning the thickness coordinate. Highlighting the defects associated with HSDTs, Garg and Chalak [2] reported,

These theories predicted the continuous transverse shear strain variation across the thickness at interfaces with a discontinuity in the transverse shear stresses. Also,

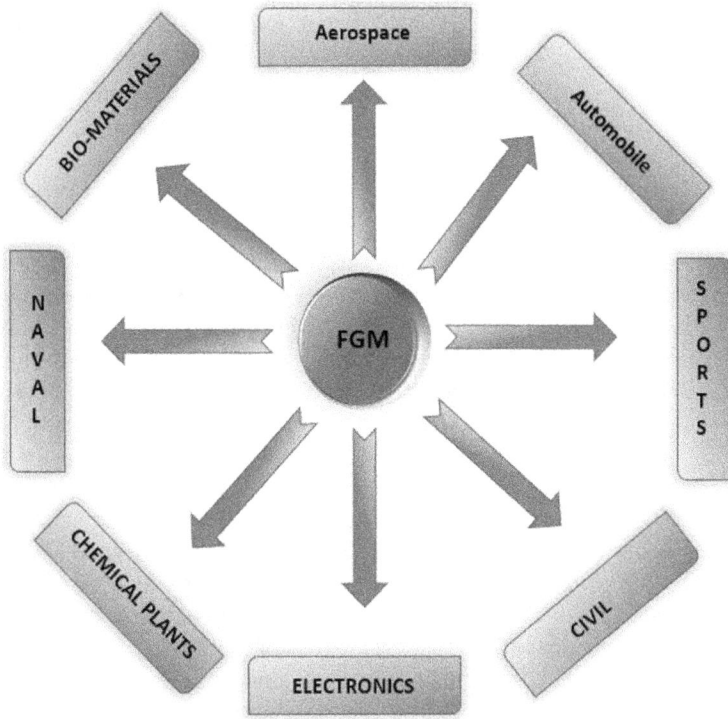

FIGURE 7.3 Applications of FGMs.

additional dependent unknowns are introduced in HSDT with each new adding power of thickness coordinate. Even these theories do not satisfy the stress-free condition at the top and the bottom face of the laminated structure.

The limitations posed by HSDT are taken care of by layer-wise theories (LWT) in which each layer is analyzed separately. Two different LWTs are widely used, namely discrete LWT and refined LWT (also called zigzag theories). In the case of discrete LWT, each layer is analyzed separately, and then the results are integrated for all layers. But this theory is computationally costly as with the increase in layers, the number of unknowns increases. Zigzag theories (ZZT) write unknowns at each layer concerning the reference layer and are computationally efficient. The most exact theory for the analysis of FGM and sandwich FGM structures is the elasticity theory. But this theory is computationally very costly. A brief summary of the theories can be found in the work reported by the authors [2, 9, 11, 23, 24].

In the present work, the authors have tried to provide an insight into the work carried out to date for the analysis of sandwich FGM structures (plate, beam, and shells) under various loading conditions.

7.2 ANALYSIS OF FGM STRUCTURE

In this section, a review of recent literature is summarized. The study is classified based on static/bending, vibration, and buckling analysis of FG and sandwich FG plates, beams, and shells.

7.2.1 BENDING STUDIES

Zhang and Zhou [25] carried out bending analysis of FGM plates using the neutral surface method. Dai et al. [26] employed the weak form meshfree Galerkin method and the least square error method using FSDT for thermo-mechanical analysis of FGM plates. Thermo-mechanical-based bending analysis of FG plates resting upon nonlinear foundations has been carried out by Yu et al. [27] and also who proposed reported isogeometric-based FSDT for the nonlinear analysis of FG plates. Nguyen et al. [28] reported the analysis of FG and sandwich FGM beams using FSDT. Sepahi et al. [29] reported differential quadrature method-based FSDT. Tornabene and Reddy [30] proposed generalized differential quadrature-based FSDT for the static analysis of FGM shells. Doubly curved FGM shells resting upon elastic foundations were analyzed by Kiani et al. [31] using Navier's solution based on FSDT and Sander's shell theory. Zidi et al. [32] employed a four-variable-based theory for hygro-thermo-mechanical-based bending analysis of FGM plates.

Using sinusoidal shear deformation theory, Zenkour [33] carried out bending analysis of FG plates and sandwich FG plates. Later, Zenkour and Alghamdi [34] carried out thermal-based bending analysis of FG plates and sandwich FG plates using the same theory considering the transverse displacement field [33] and also considering the constant transverse displacement field [35, 36]. With the help of the energy concept, Khabbaz et al. [37] carried out an analysis of FGM plates in the nonlinear range. Ben-Oumrane et al. [38] carried out a comparative study between various shear deformation theories for the bending analysis of sigmoidal FGM beams. Nguyen and Nguyen [39] proposed analytical solutions for the bending of FG beams and sandwich FG beams using inverse hyperbolic shear deformation theory. Galerkin-based solutions were proposed by Apetre et al. [40] for sandwich FG beams under bending conditions. Taj et al. [41] reported bending analysis of skew sandwich FG plates and non-skew FG plates [42] using HSDT-based finite element analysis. Taj and Chakrabarti [43] also analyzed the bending behavior of FG shells using the HSDT accounting transverse displacement field.

Talha and Singh [44] used FE-based HSDT for the bending behavior of FGM plates. Several works have been published by Tornabene and his co-authors [45–47] on the free vibration analysis of FGM shells with different geometries. Using stress equilibrium equations, Li et al. [48] carried out analysis of sandwich FGM beams using mixed finite element analysis. Zenkour [49] carried out bending analysis of FG plates and sandwich FGM plates with porosities using HSDT. Koutoati et al. [50] carried out bending analysis of FGM beams and sandwich FGM beams using finite element-based HSDT. Benbakhti et al. [51] proposed Navier's solution-based HSDT (containing five variables) for thermo-mechanical-based bending analysis of

FG plates. Golmakani and Kadkhodayan [52] carried out bending analysis of annular FGM plates using the dynamic relaxation method in nonlinear range with HSDT and FSDT. Later, the methodology was extended for the thermo-mechanical-based bending analysis of FGM plates [53]. Kumar et al. [54] proposed Wendland radial basic function (RBF) meshfree method-based HSDT for the bending analysis of FGM plates under patch load. Singh and Harsha [55] carried out bending analysis of S-FGM plates using Galerkin Vlasov's method and HSDT.

Neves et al. [56–58] carried out a bending study on power-law-based sandwich FGM plates using hyperbolic zigzag theory. Fares et al. [59] carried out the analysis of FG shells using LWT. The analysis of sandwich FGM plates has been reported by Pandey and Pradyumna [60, 61] using finite element-based LWT. Iurlaro et al. [62] carried out bending analysis of FGM and sandwich FGM plates using HOZT. Garg et al. [12] carried out a bending-based comparative study on sandwich FGM beams made up of different kinds of material homogenization rules. Garg et al. [63] carried out finite element-based bending analysis of power-law and exponential sandwich FGM plates using HOZT.

Filippi et al. [64] employed finite element-based Carrera's unified formulation for the bending analysis of FGM beams. The authors also carried out a comparative study of the efficiency of various theories in predicting the behavior of FGM beams.

Sankar and his co-authors carried out elasticity-based bending analysis of FGM beams and sandwich FGM beams [65–67]. Kashtalyan and Menshykova [68] proposed elasticity-based bending solutions for exponential FG and sandwich FG plates. Liu et al. [69] published 3D elasticity solutions for the bending analysis of circular FGM plates.

7.2.2 Vibration Studies

Yin et al. [70] carried out a free vibration analysis of thin FG plates using classical plate theory and physical neutral surface. Asanjarani et al. [71] analyzed 2D FGM conical shells resting on elastic foundations using FSDT. Nguyen et al. [28] carried out a vibration analysis of FG and sandwich FGM beams using FSDT. Bourada et al. [72] carried out stability analysis of symmetric FGM plates using four-variables-based SDT, which does not require a shear correction factor. Civalek and Baltacıoglu [73] reported a free vibration study of the annular sector and sector plates using FSDT and Love's conical shell theory. The authors used the method of harmonic differential quadrature (HDQ) and discrete singular convolution (DSC). Katili et al. [74] carried out a free vibration analysis of FGM beams using the Timoshenko beam theory. Tabatabaei and Fattahi [75] carried out finite element-based modal analysis of FGM plates with different boundary conditions.

Zenkour [76] carried out free vibration analysis of FG and sandwich FG plates using sinusoidal shear deformation theory. Murin et al. [77] worked out exact solutions for free vibration problems of FG beams. Neves et al. [78] carried out a radial basis function-based HSDT for the free vibration analysis of sandwich FGM plates. Tran et al. [79] studied the free vibration behavior of cracked FGM plates using an extended isogeometric approach based on HSDT. Nguyen et al. [80] proposed an analytical solution-based HSDT for the vibration analysis of FGM beams. Several works are reported by Tornabene and his co-authors [81–84] for free vibration analysis of FGM shells having different geometries. Osofero et al. [85] carried out a free vibration analysis of FG and sandwich FGM beams using exponential, sinusoidal, and hyperbolic shear deformation theories. Using inverse hyperbolic shear deformation theory, Nguyen and Nguyen [39] proposed analytical solutions for free vibration analysis of FG and sandwich FG beams. Shen et al. [86] carried out a free vibration analysis of doubly curved FGM shells under thermal conditions. Do and Lee [87] carried out a free vibration analysis of FGM plates with cut-outs using the isogeometric method applied in the framework of HSDT. Lal and Saini [88–91] carried out a thermal-based vibration analysis of FGM plates using a generalized differential quadrature method. Koutoati et al. [50] carried out free vibration analysis of FGM beams and sandwich FGM beams using finite element-based HSDT. Van et al. [92, 93] carried out the transient analysis of FGM plates under mechanical and hygro-thermo-mechanical loadings. Baghlani et al. [94] carried out free vibration analysis of FGM cylindrical shells resting upon an elastic foundation under a thermal environment. The study also included the effects of stiffeners, fluid load, etc. Aris and Ahmadi [95] carried out the nonlinear vibration analysis of cylindrical FGM shells. Using an isogeometric B-spline finite strip method, Shahmohammadi et al. [96] reported a free vibration study of sandwich FGM shells. Babaei et al. [97] proposed closed-form solutions based on large vibration analysis of shallow FGM arches resting upon elastic foundations.

7.2.3 BUCKLING STUDIES

Khoa and Tung [98] carried out a buckling analysis of sandwich FGM spherical shallow shells under a thermal environment. Hajlaoui et al. [99, 100] reported buckling behavior of FGM plates and shells using shear correction factor-based FSDT along with the finite element method. Akbari and Asanjarani [101] carried out mechanical and thermal buckling analysis of 2D-FG circular plates. Chan and his co-authors [102–105] carried out buckling analysis of stiffened sandwich FGM conical shells resting upon an elastic foundation using FSDT. Earlier, a similar methodology was used by Cong and Duc [106] for thermal-based buckling analysis of stiffened sigmoidal FGM plates. Trabelsi et al. [107] proposed a modified form of finite element-based FSDT for buckling analysis of FGM plates and cylindrical shells. Using Euler–Bernoulli beam theory, Kiss [108] carried out a stability analysis of FGM arches.

Zenkour [76] carried out a buckling analysis of FG and sandwich FG plates using sinusoidal shear deformation theory. Zenkour and Sobhy [109] extended the same theory [76] for buckling FGM plates' analysis under thermal loadings. Taj and Chakrabarti [110] carried out buckling analysis of skew FG plates using finite element-based HSDT. Nguyen et al. [80] proposed an analytical solution based on HSDT for buckling analysis of FGM beams. Osofero et al. [85] carried out a buckling analysis of FG and sandwich FGM beams using exponential, sinusoidal, and hyperbolic shear deformation theories. Nguyen and Nguyen [39] proposed analytical solutions for buckling analysis of FG and sandwich FG beams using inverse hyperbolic shear deformation theory. Yang et al. [111] carried out a buckling analysis of FGM plates with internal defects and cut-outs using extended isogeometric analysis based on PHT-splines and finite cell method-based HSDT. Zghal et al. [112] carried out a buckling analysis of FGM plates and shells. Zaoui et al. [113] carried out the dynamic analysis of FGM plates using analytical solution-based HSDT containing fewer variables than the conventional HSDTs. Do et al. [114] carried out a nonlinear buckling analysis of FGM plates considering different thermal conditions. Neves et al. [115] carried out a buckling analysis of sandwich FGM plates using HOZT.

7.3 DISCUSSION

Garg et al. [12, 63], in their work, have shown that the inclusion of zigzag effects in the transverse displacement field helps in predicting the transverse stresses efficiently. The inclusion of zigzag effects is alone not sufficient for describing the behavior of FGM structures efficiently. Transverse normal strains or transverse displacement fields must be included during formulation [116]. The sandwich FGM structural element's behavior with symmetric thickness scheme is different from the structural element having an unsymmetric thickness scheme [12].

Burlayenko and Sadowski [117] reported that stresses have a nonlinear pattern across sandwich FGM plate thickness. Authors also reported that

> The volume fraction of ceramic on the top surface is an important factor for dynamic design of FGM sandwich plates. Less is the ceramic concentration on the top surface, the less sensitive are the frequencies to increasing the metallic phase in the sandwich core.

The nonlinear variation increases with an increase in the metallic phase in the plate for plates made with the power law.

The use of post-processing techniques such as equilibrium conditions and least-square error methods adds to the computational efforts, which are mainly used to describe the transverse shear stress-free condition at the top and bottom surfaces of the FG structural element along with the continuity condition at the interfaces. When face-to-core stiffness thickness is not high, FSDT and HSDT can predict sandwich FGM structures' behavior with good accuracy. But, as the ratio increases, both theories are unable to give good results. Thickness-stretching and extensional dominant deformations were also observed in mode shapes along with flexural dominant deformations for thick FG and sandwich FGM plates.

7.4 CONCLUSION AND FUTURE SCOPE

The following are the important points noted down during the literature survey of the recent work carried out on the bending, vibration, and buckling analysis of FG and sandwich FG beams, plates, and shells:

1. The power-law is the most widely used law for the analysis of FG and sandwich FG structural elements.

2. Navier's solution is the most widely used analytic method for predicting the behavior of FG structures, followed by the finite element method.
3. Some new methods, such as the radial point interpolation method, Galerkin Vlasov's method, peridynamic differential operator, etc., are successfully applied for predicting the behavior of FG structures.
4. The behavior of sandwich FG structures is different from that of FG structures. Symmetric sandwich FG structures give symmetric stress distribution across the thickness.
5. The application of HOZT for the analysis of FG beams and sandwich FG beams and shells is not fully explored.
6. Analysis of sandwich FG shells with complex configurations has not been explored fully.
7. Limited studies are reported in the literature on the comparative studies on sandwich FG structural elements under different loading conditions.

ACKNOWLEDGMENT

The authors thank Rahul Sakla (research scholar, Department of Chemistry, National Institute of Technology Kurukshetra) for plotting Figure 7.1.

REFERENCES

1. Liew KM, Zhao X, Ferreira AJM. A review of meshless methods for laminated and functionally graded plates and shells. *Compos Struct* 2011;93:2031–41. doi:10.1016/j.compstruct.2011.02.018.
2. Garg A, Chalak HD. A review on analysis of laminated composite and sandwich structures under hygrothermal conditions. *Thin-Walled Struct* 2019;142:205–26. doi:10.1016/j.tws.2019.05.005.
3. Caliri MF, Ferreira AJM, Tita V. A review on plate and shell theories for laminated and sandwich structures highlighting the finite element method. *Compos Struct* 2016;156:63–77. doi:10.1016/j.compstruct.2016.02.036.
4. Sayyad AS, Ghugal YM. On the free vibration analysis of laminated composite and sandwich plates: A review of recent literature with some numerical results. *Compos Struct* 2015;129:177–201. doi:10.1016/j.compstruct.2015.04.007.
5. Mackerle J. Finite element analyses of sandwich structures: A bibliography (1980–2001). *Eng Computation* 2002;19(2):206–245. doi:10.1108/02644400210419067.
6. Noor AK, Burton WS. Computational models for high-temperature multilayered composite plates and shells. *Appl Mech Rev* 1992;45:419–46.
7. Patni M, Minera S, Groh RMJ, Pirrera A, Weaver PM. Three-dimensional stress analysis for laminated composite and sandwich structures. *Compos B Eng* 2018;155:299–328. doi:10.1016/j.compositesb.2018.08.127.
8. Garg A, Chalak H. Analysis of non-skew and skew laminated composite and sandwich plates under hygro-thermo-mechanical conditions including transverse stress variations. *J Sandw Struct Mater* 2020:109963622093278. doi:10.1177/1099636220932782.
9. Garg A, Chalak H. Novel higher-order zigzag theory for analysis of laminated sandwich beams. *Proc Inst Mech Eng L* 2021: 235(1): 176–94. doi:10.1177/1464420720957045.
10. Ghatage PS, Kar VR, Sudhagar PE. On the numerical modelling and analysis of multi-directional functionally graded composite structures: A review. *Compos Struct* 2020;236:111837. doi:10.1016/j.compstruct.2019.111837.

11. Jha DK, Kant T, Singh RK. A critical review of recent research on functionally graded plates. *Compos Struct* 2013;96:833–49. doi:10.1016/j.compstruct.2012.09.001.

12. Garg A, Chalak HD, Chakrabarti A. Comparative study on the bending of sandwich FGM beams made up of different material variation laws using refined layerwise theory. *Mech Mater* 2020;151:103634. doi:10.1016/j.mechmat.2020.103634.

13. Saleh B, Jiang J, Fathi R, Al-hababi T, Xu Q, Wang L, et al. 30 years of functionally graded materials: An overview of manufacturing methods, applications and future challenges. *Compos B Eng* 2020;201:108376. doi:10.1016/j.compositesb.2020.108376.

14. Swaminathan K, Naveenkumar DT, Zenkour AM, Carrera E. Stress, vibration and buckling analyses of FGM plates-A state-of-the-art review. *Compos Struct* 2015;120:10–31. doi:10.1016/j.compstruct.2014.09.070.

15. Sofiyev AH. Review of research on the vibration and buckling of the FGM conical shells. *Compos Struct* 2019;211:301–17. doi:10.1016/j.compstruct.2018.12.047.

16. Garg A, Belarbi M-O, Chalak HD, Chakrabarti A. A review of the analysis of sandwich FGM structures. *Compos Struct* 2021;258:113427. doi:10.1016/j.compstruct.2020.113427.

17. Chi SH, Chung YL. Mechanical behavior of functionally graded material plates under transverse load-Part I: Analysis. *Int J Solids Struct* 2006;43(13):3657–74. doi:10.1016/j.ijsolstr.2005.04.011.

18. Chi SH, Chung YL. Mechanical behavior of functionally graded material plates under transverse load-Part II: Numerical results. *Int J Solids Struct* 2006. doi:10.1016/j.ijsolstr.2005.04.010.

19. Pai PF. A new look at shear correction factors and warping functions of anisotropic laminates. *Int J Solids Struct* 1995;32:2295–313. doi:10.1016/0020-7683(94)00258-X.

20. Huang NN. Influence of shear correction factors in the higher order shear deformation laminated shell theory. *Int J Solids Struct* 1994;31:1263–77. doi:10.1016/0020-7683(94)90120-1.

21. Birman V, Bert CW. On the choice of shear correction factor in sandwich structures. *J Sandw Struct Mater* 2002;4:83–95. doi:10.1177/1099636202004001180.

22. Nguyen TK, Sab K, Bonnet G. Shear correction factors for functionally graded plates. *Mech Adv Mater Struct* 2007;14(8):567–75. doi:10.1080/15376490701672575.

23. Carrera E. An assessment of mixed and classical theories for the thermal stress analysis of orthotropic multilayered plates. *J Therm Stress* 2000;23:797–831. doi:10.1080/014957300750040096.

24. Carrera E. Historical review of Zig-Zag theories for multilayered plates and shells. *Appl Mech Rev* 2003;56:287. doi:10.1115/1.1557614.

25. Zhang DG, Zhou YH. A theoretical analysis of FGM thin plates based on physical neutral surface. *Comput Mater Sci* 2008;44(2):716–20. doi:10.1016/j.commatsci.2008.05.016.

26. Dai KY, Liu GR, Han X, Lim KM. Thermomechanical analysis of functionally graded material (FGM) plates using element-free Galerkin method. *Comput Struct*, 2005;83:1487–1502. doi:10.1016/j.compstruc.2004.09.020.

27. Yu TT, Yin S, Bui TQ, Hirose S. A simple FSDT-based isogeometric analysis for geometrically nonlinear analysis of functionally graded plates. *Finite Elem Anal Des* 2015;96:1–10. doi:10.1016/j.finel.2014.11.003.

28. Nguyen TK, Vo TP, Thai HT. Static and free vibration of axially loaded functionally graded beams based on the first-order shear deformation theory. *Compos B Eng* 2013;55:147–57. doi:10.1016/j.compositesb.2013.06.011.

29. Sepahi O, Forouzan MR, Malekzadeh P. Large deflection analysis of thermo-mechanical loaded annular FGM plates on nonlinear elastic foundation via DQM. *Compos Struct* 2010;92:2369–78. doi:10.1016/j.compstruct.2010.03.011.

30. Tornabene F, Reddy JN. FGM and laminated doubly-curved and degenerate shells resting on nonlinear elastic foundations: A GDQ solution for static analysis with a posteriori stress and strain recovery. *J Indian Inst Sci* 2013;93(4):635–88.

31. Kiani Y, Akbarzadeh AH, Chen ZT, Eslami MR. Static and dynamic analysis of an FGM doubly curved panel resting on the Pasternak-type elastic foundation. *Compos Struct* 2012;94(8):2474–84. doi:10.1016/j.compstruct.2012.02.028.

32. Zidi M, Tounsi A, Houari MSA, Adda Bedia EA, Anwar Bég O. Bending analysis of FGM plates under hygro-thermo-mechanical loading using a four variable refined plate theory. *Aerosp Sci Technol* 2014;34:24–34. doi:10.1016/j.ast.2014.02.001.

33. Zenkour AM. A comprehensive analysis of functionally graded sandwich plates: Part 1—Deflection and stresses. *Int J Solids Struct* 2005;42:5224–42. doi:10.1016/j.ijsolstr.2005.02.015.

34. Zenkour AM, Alghamdi NA. Thermoelastic bending analysis of functionally graded sandwich plates. *J Mater Sci* 2008;43:2574–89. doi:10.1007/s10853-008-2476-6.

35. Zenkour AM, Alghamdi NA. Thermomechanical bending response of functionally graded nonsymmetric sandwich plates. *J Sandw Struct Mater* 2010;12:7–46. doi:10.1177/1099636209102264.

36. Zenkour AM, Alghamdi NA. Bending analysis of functionally graded sandwich plates under the effect of mechanical and thermal loads. *Mech Adv Mater Struct* 2010;17:419–32. doi:10.1080/15376494.2010.483323.

37. Khabbaz RS, Manshadi BD, Abedian A. Nonlinear analysis of FGM plates under pressure loads using the higher-order shear deformation theories. *Compos Struct* 2009;89(3):333–44. doi:10.1016/j.compstruct.2008.06.009.

38. Ben-Oumrane S, Abedlouahed T, Ismail M, Mohamed BB, Mustapha M, El Abbas AB. A theoretical analysis of flexional bending of Al/Al2O3 S-FGM thick beams. *Comput Mater Sci* 2009;44(4):1344–50. doi:10.1016/j.commatsci.2008.09.001.

39. Nguyen T-K, Nguyen B-D. A new higher-order shear deformation theory for static, buckling and free vibration analysis of functionally graded sandwich beams. *J Sandw Struct Mater* 2015;17:613–31. doi:10.1177/1099636215589237.

40. Apetre NA, Sankar BV., Ambur DR. Analytical modeling of sandwich beams with functionally graded core. *J Sandw Struct Mater* 2008;10:53–74. doi:10.1177/1099636207081111.

41. Taj MG, Chakrabarti A, Talha M. Bending analysis of functionally graded skew sandwich plates with through-the thickness displacement variations. *J Sandw Struct Mater* 2014;16:210–48. doi:10.1177/1099636213512499.

42. Gulshan Taj MNA, Chakrabarti A, Sheikh AH. Analysis of functionally graded plates using higher order shear deformation theory. *Appl Math Model* 2013;37:8484–94. doi:10.1016/j.apm.2013.03.058.

43. Taj G, Chakrabarti A. Modeling of functionally graded sandwich shells accounting for variation in transverse displacement. *Mech Adv Mater Struct* 2017;24(6):509–23. doi:10.1080/15376494.2016.1145773.

44. Talha M, Singh BN. Static response and free vibration analysis of FGM plates using higher order shear deformation theory. *Appl Math Model* 2010;34:3991–4011. doi:10.1016/j.apm.2010.03.034.

45. Tornabene F, Fantuzzi N, Viola E, Batra RC. Stress and strain recovery for functionally graded free-form and doubly-curved sandwich shells using higher-order equivalent single layer theory. *Compos Struct* 2015;119:67–89. doi:10.1016/j.compstruct.2014.08.005.

46. Tornabene F, Ceruti A. Mixed static and dynamic optimization of four-parameter functionally graded completely doubly curved and degenerate shells and panels using GDQ method. *Math Probl Eng* 2013;867079:1–33. doi:10.1155/2013/867079.

47. Tornabene F, Viola E. Static analysis of functionally graded doubly-curved shells and panels of revolution. *Meccanica* 2013;48:901–30. doi:10.1007/s11012-012-9643-1.

48. Li W, Ma H, Gao W. A higher-order shear deformable mixed beam element model for accurate analysis of functionally graded sandwich beams. *Compos Struct* 2019;221:110830. doi:10.1016/j.compstruct.2019.04.002.

49. Zenkour AM. A quasi-3D refined theory for functionally graded single-layered and sandwich plates with porosities. *Compos Struct* 2018;201:38–48. doi:10.1016/j.compstruct.2018.05.147.

50. Koutoati K, Mohri F, Daya EM. Finite element approach of axial bending coupling on static and vibration behaviors of functionally graded material sandwich beams. *Mech Adv Mater Struct* 2019:1–17. doi:10.1080/15376494.2019.1685144.

51. Benbakhti A, Bouiadjra MB, Retiel N, Tounsi A. A new five unknown quasi-3D type HSDT for thermomechanical bending analysis of FGM sandwich plates. *Steel Compos Struct* 2016;22(5):975–99. doi:10.12989/scs.2016.22.5.975.

52. Golmakani ME, Kadkhodayan M. Nonlinear bending analysis of annular FGM plates using higher-order shear deformation plate theories. *Compos Struct* 2011;93(2):973–82. doi:10.1016/j.compstruct.2010.06.024.

53. Golmakani ME, Kadkhodayan M. Large deflection analysis of circular and annular FGM plates under thermo-mechanical loadings with temperature-dependent properties. *Compos B Eng* 2011;42(4):614–25. doi:10.1016/j.compositesb.2011.02.018.

54. Kumar R, Lal A, Singh BN, Singh J. New transverse shear deformation theory for bending analysis of FGM plate under patch load. *Compos Struct* 2019;208:91–100. doi:10.1016/j.compstruct.2018.10.014.

55. Singh SJ, Harsha SP. Thermo-mechanical analysis of porous sandwich S-FGM plate for different boundary conditions using Galerkin Vlasov's method: A semi-analytical approach. *Thin-Walled Struct* 2020;150:106668. doi:10.1016/j.tws.2020.106668.

56. Neves AMA, Ferreira AJM, Carrera E, Cinefra M, Jorge RMN, Soares CMM. Static analysis of functionally graded sandwich plates according to a hyperbolic theory considering Zig-Zag and warping effects. *Adv Eng Softw* 2012;52:30–43. doi:10.1016/j.advengsoft.2012.05.005.

57. Neves AMA, Ferreira AJM, Carrera E, Roque CMC, Cinefra M, Jorge RMN, et al. A quasi-3D sinusoidal shear deformation theory for the static and free vibration analysis of functionally graded plates. *Compos B Eng* 2012;43:711–25. doi:10.1016/j.compositesb.2011.08.009.

58. Neves AMA, Ferreira AJM, Carrera E, Cinefra M, Roque CMC, Jorge RMN, et al. Static, free vibration and buckling analysis of isotropic and sandwich functionally graded plates using a quasi-3D higher-order shear deformation theory and a meshless technique. *Compos B Eng* 2013;44:657–74. doi:10.1016/j.compositesb.2012.01.089.

59. Fares ME, Elmarghany MK, Atta D, Salem MG. Bending and free vibration of multilayered functionally graded doubly curved shells by an improved layerwise theory. *Compos B Eng* 2018;154:272–84. doi:10.1016/j.compositesb.2018.07.038.

60. Pandey S, Pradyumna S. Analysis of functionally graded sandwich plates using a higher-order layerwise theory. *Compos Part B* 2018;153:325–36. doi:10.1016/j.compositesb.2018.08.121.

61. Pandey S, Pradyumna S. Free vibration of functionally graded sandwich plates in thermal environment using a layerwise theory. *Eur J Mech A Solids* 2015;51:55–66. doi:10.1016/j.euromechsol.2014.12.001.

62. Iurlaro L, Gherlone M, Di Sciuva M. Bending and free vibration analysis of functionally graded sandwich plates using the Refined Zigzag Theory. *J Sandw Struct Mater* 2014;16:669–99. doi:10.1177/1099636214548618.

63. Garg A, Chalak HD, Chakrabarti A. Bending analysis of functionally graded sandwich plates using HOZT including transverse displacement effects. *Mech Based Des Struct Mach* 2020:;0(0)1–15. doi:10.1080/15397734.2020.1814157.

64. Filippi M, Carrera E, Zenkour AM. Static analyses of FGM beams by various theories and finite elements. *Compos B Eng* 2015;72:1–9. doi:10.1016/j.compositesb.2014.12.004.

65. Venkataraman S, Sankar BV. Elasticity Solution for Stresses in a Sandwich Beam with Functionally Graded Core. *AIAA J* 2003;41:2501–5. doi:10.2514/2.6853.

66. Venkataraman S, Sankar B. Analysis of sandwich beams with functionally graded core. 19th AIAA Appl. Aerodyn. Conf., Reston, Virigina: American Institute of Aeronautics and Astronautics; 2001;1281:1–8. doi:10.2514/6.2001-1281.

67. Sankar BV. An elasticity solution for functionally graded beams. *Compos Sci Technol* 2001;61(5):689–96. doi:10.1016/S0266-3538(01)00007-0.

68. Kashtalyan M, Menshykova M. Three-dimensional elasticity solution for sandwich panels with a functionally graded core. *Compos Struct* 2009;87:36–43. doi:10.1016/j.compstruct.2007.12.003.

69. Liu NW, Sun YL, Chen WQ, Yang B, Zhu J. 3D elasticity solutions for stress field analysis of FGM circular plates subject to concentrated edge forces and couples. *Acta Mech* 2019;230:2655–68. doi:10.1007/s00707-019-02412-z.

70. Yin S, Yu T, Liu P. Free vibration analyses of FGM thin plates by isogeometric analysis based on classical plate theory and physical neutral surface. *Adv Mech Eng* 2013. doi:10.1155/2013/634584.

71. Asanjarani A, Satouri S, Alizadeh A, Kargarnovin MH. Free vibration analysis of 2D-FGM truncated conical shell resting on Winkler-Pasternak foundations based on FSDT. *Proc Inst Mech Eng C* 2015;229(5):818–39. doi:10.1177/0954406214539472.

72. Bourada F, Amara K, Bousahla AA, Tounsi A, Mahmoud SR. A novel refined plate theory for stability analysis of hybrid and symmetric S-FGM plates. *Struct Eng Mech* 2018;68(6):661–75. doi:10.12989/sem.2018.68.6.661.

73. Civalek Ö, Baltacıoglu AK. Free vibration analysis of laminated and FGM composite annular sector plates. *Compos B Eng* 2019;157:182–94. doi:10.1016/j.compositesb.2018.08.101.

74. Katili I, Syahril T, Katili AM. Static and free vibration analysis of FGM beam based on unified and integrated of Timoshenko's theory. *Compos Struct* 2020;242:112–130. doi:10.1016/j.compstruct.2020.112130.

75. Tabatabaei SJS, Fattahi AM. A finite element method for modal analysis of FGM plates. *Mech Based Des Struct Mach* 2020;(0):1–12. doi:10.1080/15397734.2020.1744004.

76. Zenkour AM. A comprehensive analysis of functionally graded sandwich plates: Part 2—Buckling and free vibration. *Int J Solids Struct* 2005;42:5243–58. doi:10.1016/j.ijsolstr.2005.02.016.

77. Murín J, Aminbaghai M, Kutiš V. Exact solution of the bending vibration problem of FGM beams with variation of material properties. *Eng Struct* 2010. doi:10.1016/j.engstruct.2010.02.010.

78. Neves AMA, Ferreira AJM, Carrera E, Cinefra M, Roque CMC, Jorge RMN, et al. Free vibration analysis of functionally graded shells by a higher-order shear deformation theory and radial basis functions collocation, accounting for through-the-thickness deformations. *Eur J Mech A Solids* 2013. doi:10.1016/j.euromechsol.2012.05.005.

79. Tran L V., Ly HA, Lee J, Wahab MA, Nguyen-Xuan H. Vibration analysis of cracked FGM plates using higher-order shear deformation theory and extended isogeometric approach. *Int J Mech Sci* 2015. doi:10.1016/j.ijmecsci.2015.03.003.

80. Nguyen T-K, Vo TP, Nguyen B-D, Lee J. An analytical solution for buckling and vibration analysis of functionally graded sandwich beams using a quasi-3D shear deformation theory. *Compos Struct* 2016;156:238–52. doi:10.1016/j.compstruct.2015.11.074.

81. Tornabene F. Free vibration analysis of functionally graded conical, cylindrical shell and annular plate structures with a four-parameter power-law distribution. *Comput Methods Appl Mech Eng* 2009;198:2911–35. doi:10.1016/j.cma.2009.04.011.

82. Tornabene F, Liverani A, Caligiana G. FGM and laminated doubly curved shells and panels of revolution with a free-form meridian: A 2-D GDQ solution for free vibrations. *Int J Mech Sci* 2011;53(6):446–70. doi:10.1016/j.ijmecsci.2011.03.007.

83. Tornabene F. 2-D GDQ solution for free vibrations of anisotropic doubly-curved shells and panels of revolution. *Compos Struct* 2011;93(7):1854–76. doi:10.1016/j.compstruct.2011.02.006.

84. Tornabene F. Free vibrations of laminated composite doubly-curved shells and panels of revolution via the GDQ method. *Comput Methods Appl Mech Eng* 2011;200 (9-12):931–52. doi:10.1016/j.cma.2010.11.017.

85. Osofero AI, Vo TP, Nguyen TK, Lee J. Analytical solution for vibration and buckling of functionally graded sandwich beams using various quasi-3D theories. *J Sandw Struct Mater* 2016;18:3–29. doi:10.1177/1099636215582217.

86. Shen HS, Chen X, Guo L, Wu L, Huang XL. Nonlinear vibration of FGM doubly curved panels resting on elastic foundations in thermal environments. *Aerosp Sci Technol* 2015;47:434–46. doi:10.1016/j.ast.2015.10.011.

87. Van Do VN, Lee CH. Free vibration analysis of FGM plates with complex cutouts by using quasi-3D isogeometric approach. *Int J Mech Sci* 2019;159:213–33. doi:10.1016/j.ijmecsci.2019.05.034.

88. Lal R, Saini R. Vibration analysis of FGM circular plates under non-linear temperature variation using generalized differential quadrature rule. *Appl Acoust* 2020;158:107027. doi:10.1016/j.apacoust.2019.107027.

89. Lal R, Saini R. Vibration analysis of functionally graded circular plates of variable thickness under thermal environment by generalized differential quadrature method. *JVC/J Vib Control* 2020;26(1-2):73–87. doi:10.1177/1077546319876389.

90. Lal R, Saini R. On the high-temperature free vibration analysis of elastically supported functionally graded material plates under mechanical in-plane force via GDQR. *J Dyn Syst Meas Control Trans ASME* 2019;141(10):101003. doi:10.1115/1.4043489.

91. Lal R, Saini R. Thermal effect on radially symmetric vibrations of temperature-dependent FGM circular plates with nonlinear thickness variation. *Mater Res Express* 2019;6:0865f1. doi:10.1088/2053-1591/ab24ee.

92. Phung-Van P, Thai CH, Ferreira AJM, Rabczuk T. Isogeometric nonlinear transient analysis of porous FGM plates subjected to hygro-thermo-mechanical loads. *Thin-Walled Struct* 2020;148:106497. doi:10.1016/j.tws.2019.106497.

93. Phung-Van P, Tran LV., Ferreira AJM, Nguyen-Xuan H, Abdel-Wahab M. Nonlinear transient isogeometric analysis of smart piezoelectric functionally graded material plates based on generalized shear deformation theory under thermo-electro-mechanical loads. *Nonlinear Dyn* 2017;87:879–94. doi:10.1007/s11071-016-3085-6.

94. Baghlani A, Khayat M, Dehghan SM. Free vibration analysis of FGM cylindrical shells surrounded by Pasternak elastic foundation in thermal environment considering fluid-structure interaction. *Appl Math Model* 2020;78:550–75. doi:10.1016/j.apm.2019.10.023.

95. Aris H, Ahmadi H. Nonlinear vibration analysis of FGM truncated conical shells subjected to harmonic excitation in thermal environment. *Mech Res Commun* 2020;104:103499. doi:10.1016/j.mechrescom.2020.103499.

96. Shahmohammadi MA, Azhari M, Saadatpour MM. Free vibration analysis of sandwich FGM shells using isogeometric B-spline finite strip method. *Steel Compos Struct* 2020;34(3):361–76. doi:10.12989/scs.2020.34.3.361.

97. Babaei H, Kiani Y, Eslami MR. Large amplitude free vibration analysis of shear deformable FGM shallow arches on nonlinear elastic foundation. *Thin-Walled Struct* 2019;144:106237. doi:10.1016/j.tws.2019.106237.

98. Khoa NM, Van Tung H. Nonlinear thermo-mechanical stability of shear deformable FGM sandwich shallow spherical shells with tangential edge constraints. *Viet J Mech* 2017;39(4):351–64. doi:10.15625/0866-7136/9810.

99. Hajlaoui A, Triki E, Frikha A, Wali M, Dammak F. Nonlinear dynamics analysis of FGM shell structures with a higher order shear strain enhanced solid-shell element. *Lat Am J Solids Struct* 2017;14:72–91. doi:10.1590/1679-78253323.

100. Hajlaoui A, Jarraya A, El Bikri K, Dammak F. Buckling analysis of functionally graded materials structures with enhanced solid-shell elements and transverse shear correction. *Compos Struct* 2015;132:87–97. doi:10.1016/j.compstruct.2015.04.059.

101. Akbari P, Asanjarani A. Semi-analytical mechanical and thermal buckling analyses of 2D-FGM circular plates based on the FSDT. *Mech Adv Mater Struct* 2019;26(9):753–64. doi:10.1080/15376494.2017.1410913.

102. Duc ND, Seung-Eock K, Chan DQ. Thermal buckling analysis of FGM sandwich truncated conical shells reinforced by FGM stiffeners resting on elastic foundations using FSDT. *J Therm Stress* 2018;41:331–65. doi:10.1080/01495739.2017.1398623.

103. Van Dung D, Chan DQ. Analytical investigation on mechanical buckling of FGM truncated conical shells reinforced by orthogonal stiffeners based on FSDT. *Compos Struct* 2017;159:827–41. doi:10.1016/j.compstruct.2016.10.006.

104. Chan DQ, van Dung D, Hoa LK. Thermal buckling analysis of stiffened FGM truncated conical shells resting on elastic foundations using FSDT. *Acta Mech* 2018;229:1–29. doi:10.1007/s00707-017-2090-2.

105. Van Dung D, Nga NT. Nonlinear analysis of stability for imperfect eccentrically stiffened FGM plates under mechanical and thermal loads based on FSDT. Part 1: Governing equations establishment. *Viet J Mech* 2015;37(3):187–204. doi:10.15625/0866-7136/37/3/5884.

106. Cong PH, Duc ND. Thermal stability analysis of eccentrically stiffened sigmoid-FGM plate with metal -ceramic -metal layers based on FSDT. *Cogent Eng* 2016;3(1):1182098. doi:10.1080/23311916.2016.1182098.

107. Trabelsi S, Frikha A, Zghal S, Dammak F. A modified FSDT-based four nodes finite shell element for thermal buckling analysis of functionally graded plates and cylindrical shells. *Eng Struct* 2019;178:444–59. doi:10.1016/j.engstruct.2018.10.047.

108. Kiss LP. Nonlinear stability analysis of FGM shallow arches under an arbitrary concentrated radial force. *Int J Mech Mater Des* 2020;16:91–108. doi:10.1007/s10999-019-09460-2.

109. Zenkour AM, Sobhy M. Thermal buckling of various types of FGM sandwich plates. *Compos Struct* 2010;93:93–102. doi:10.1016/j.compstruct.2010.06.012.

110. Taj MNAG, Chakrabarti A. Buckling analysis of functionally graded skew plates: An efficient finite element approach. *Int J Appl Mech* 2013;5(4):1350041. doi:10.1142/S1758825113500415.

111. Yang HS, Dong CY, Qin XC, Wu YH. Vibration and buckling analyses of FGM plates with multiple internal defects using XIGA-PHT and FCM under thermal and mechanical loads. *Appl Math Model* 2020;78:433–81. doi:10.1016/j.apm.2019.10.011.

112. Zghal S, Frikha A, Dammak F. Mechanical buckling analysis of functionally graded power-based and carbon nanotubes-reinforced composite plates and curved panels. *Compos B Eng* 2018;150:165–83. doi:10.1016/j.compositesb.2018.05.037.

113. Zaoui FZ, Tounsi A, Ouinas D, Viña Olay JA. A refined HSDT for bending and dynamic analysis of FGM plates. *Struct Eng Mech* 2020;74(1):105–19. doi:10.12989/sem.2020.74.1.105.

114. Van Do VN, Ong TH, Lee CH. Isogeometric analysis for nonlinear buckling of FGM plates under various types of thermal gradients. *Thin-Walled Struct* 2019;137:448–62. doi:10.1016/j.tws.2019.01.024.

115. Neves AMA, Ferreira AJM, Carrera E, Cinefra M, Jorge RMN, Mota Soares CM, et al. Influence of zig-zag and warping effects on buckling of functionally graded sandwich plates according to sinusoidal shear deformation theories. *Mech Adv Mater Struct* 2017;24:360–76. doi:10.1080/15376494.2016.1191095.

116. Di Sciuva M, Sorrenti M. Bending and free vibration analysis of functionally graded sandwich plates: An assessment of the refined zigzag theory. *J Sandw Struct Mater* 2019;0(0): 109963621984397. doi:10.1177/1099636219843970.

117. Burlayenko VN, Sadowski T. Free vibrations and static analysis of functionally graded sandwich plates with three-dimensional finite elements. *Meccanica* 2020;55:815–32. doi:10.1007/s11012-019-01001-7.

8 Modeling and Analysis of Smart Functionally Graded Structures

Dr. Saroj Kumar Sarangi

CONTENTS

8.1 INTRODUCTION

8.1.1 FUNCTIONALLY GRADED MATERIALS AND STRUCTURES

In modern technology, material science has gained quick development, and the research is leading to the invention of new materials. In this process, functionally graded (FG) materials are considered novel materials which are basically composite materials with microscopic inhomogeneous characteristics. The gradually changing volume fraction of the ingredients of composite gives smoothly varying or graded properties along desired directions. This develops a tendency to reduce the residual and thermal stress in the overall composite system. This also reduces the stress concentrations which are found in the adjoining layers in conventional composite materials [1–3].

DOI: 10.1201/9781003097976-8

With functionally graded composite material (FGCM), two types of graded structures can be made which may be continuous graded structures and step-wise or layered structures. In continuous graded structures, the material composition and microstructure change continuously with respect to the position, whereas in a step-wise or layered structure, the microstructure and material composition change in a step-wise manner, resulting a multilayered structure with an interface lying between discrete layers. Figure 8.1 shows a schematic representation of these two types of FG structures [4].

One variety of these FG materials is fabricated by combining metal and ceramics to suit particular applications where the ceramic material gives the property of high thermal resistance and the metal part gives strength as well as toughness to the structure.

FG plates and shells are essential structural elements frequently used in leading civil as well as aerospace structures. These FG structures play a vital role when the overall system is subjected to an elevated temperature environment. The surfaces of these structures are made metal/ceramic rich and the properties of materials across the thickness are continuously varying so that the structures can better resist temperature variation, maintaining the structural strength as well as toughness. The surface rich in ceramic, having low thermal conductivity, is exposed to high temperatures, giving better thermal resistance. The metallic constituent of the composite provides toughness to the plate structure. Thin-walled plate and shell structures used in turbines, rocket heat shields, heat exchanger tubes, heat-engine components, etc., may fail from buckling and large deflections induced by thermo-mechanical loading. Such thin-walled structures can be designed by tailoring the laminated composites generally by changing the ply material, its thickness, and the stacking sequence. The abrupt property variations of adjacent layers result in higher shear stresses which may initiate imperfections like de-lamination. Such harmful effects can be reduced by continuous change of properties along the thickness of the structure.

In the recent past, there has been increasing research on the FG materials and structures due to their high performance and multifunctional role, and they are used in spacecraft, nuclear reactor components, and chemical plants [5–8]. Different applications of FG materials are available in the literature, and the material properties are mostly designed graded along the thickness of the structure. These

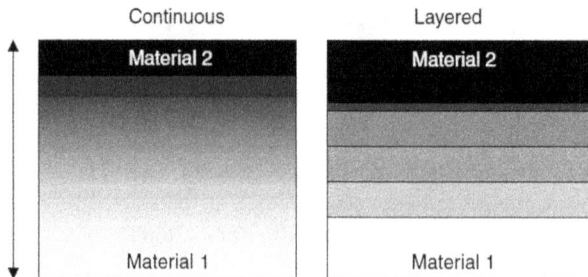

FIGURE 8.1   Representation of gradation in FG composite material.

material properties are generally found using one of the property distribution laws as explained below.

8.1.2 POWER LAW

The power law is very commonly used in designing functionally graded materials and is explained as [9]

$$P_Z = \left(P_U - P_L\right)\left(\frac{z}{h} + \frac{1}{2}\right)^k + P_L \tag{8.1}$$

in which P_Z denotes the properties such as modulus of elasticity, density, etc., for FG material. P_U and P_L denote material properties of FG structures respectively at the top-most and bottom-most surfaces, h represents the total thickness of the structure, and k represents the non-negative power-law index or material grading index which depends upon the design requirements.

8.1.3 EXPONENTIAL LAW

For FG material modeling in fracture mechanics, this particular law is more preferred and is given by [10]

$$P_Z = P_U \exp\left(-\delta\left(1 - \frac{2z}{h}\right)\right) \tag{8.2}$$

where $\delta = \frac{1}{2}\ln\left(\frac{P_U}{P_L}\right)$, and P_Z is the classic material property at a point situated at distance z measured from reference.

8.1.4 SIGMOID FUNCTION

The volume fraction of FG structures in this sigmoid function consists of two functions and can be written as [11]

$$P_z = \left(P_t - P_b\right)\left[\frac{1}{2}\left(\frac{z}{h} - \frac{1}{2}\right)^k\right] + P_b, \quad 0 \le z \le \frac{h}{2}$$

$$P_z = \left(P_t - P_b\right)\left[1 - \frac{1}{2}\left(\frac{z}{h} - \frac{1}{2}\right)^k\right] + P_b, \quad -\frac{h}{2} \le z \le 0 \tag{8.3}$$

where h and k are the FG layer thickness and material parameter respectively.

Works devoted to the study of functionally graded structures have great importance in recent decades because of their wide areas of application and are available in the literature [12–21].

8.2 SMART COMPOSITE MATERIALS

In the recent past, piezoelectric materials have been effectively used in sensors as well as actuators in smart composite structures. These monolithic materials have low control capability, the reason being the low values of their piezoelectric coefficients. Again monolithic piezoelectric materials are not compatible with curved surfaces as they are brittle in nature. It has been shown that the brittle piezoelectric fibers can be efficiently used for making polymer composites with better properties, good enough for structural applications, and perhaps, this has inspired researchers to develop piezoelectric composites. Piezoelectric composites (PZCs) are mostly prepared by reinforcing the piezoceramic fibers with conventional epoxy matrix; PZT, PZT5H, etc., are the commonly used piezoceramics. The effective material properties offered by these PZCs are better than those of the monolithic piezoelectric materials [22–24]. These piezoelectric composites possess good compatibility and strength characteristics and also they have the capability to cause anisotropic actuations. The piezoelectric composites are considered advanced smart materials which can be best suited for structural applications because of their tailor-made properties. In addition to strain and force compatibility, the actuators made of the piezoelectric composites have improved damage tolerance ability.

Initially the research work on piezoelectric composites was focused on predicting their effective properties [25–31]. Different micromechanics models have been developed to calculate the properties of the constituent materials of piezoelectric composites. Another model was developed by Chan and Unsworth [32] following a micromechanics approach to analyze piezo ceramic/polymer 1-3 composites, while Mallik and Ray [26] proposed horizontally reinforced piezoelectric composites using a strength of material approach. Smith and Auld [33] developed the properties of 1-3 piezoelectric composites for vertical reinforcement, and subsequently, Ray and Pradhan [34] developed a micromechanics model for these piezoelectric composites having vertical reinforcements.

Bent and Hagood [35] developed the active fiber composite (AFC) material to achieve in-plane actuation by the piezoelectric composites. A diagrammatic representation of active fiber composite is shown in Figure 8.2.

AFC is made of piezoceramic fibers aligned horizontally so that the fibers are length-wise poled and are enclosed in epoxy matrix and then kept between two interdigitated electrode (IDE) fingers as illustrated (Figure 8.2). These finger electrodes are lookalikes of each other, and each pattern is composed of positive as well as negative electrode fingers which are alternatively aligned. These fingers are placed transversal to the fibers and subsequently, an electric field is developed along the fibers' length. The distance (d_p) between a positive electrode and a negative electrode controls the measure of the electric field. This arrangement of electrodes in AFC renders high actuation authority in the plane along the fiber direction. AFC is one of the piezoelectric composite materials which is commercially available. An analytical model of AFC was developed by Kar-Gupta and Venkatesh [36] to analyze the electromechanical response of AFCs, while Ivanov [37] carried out finite element analysis and modeling of AFCs including damages. Recently Tripathy et al. [38]

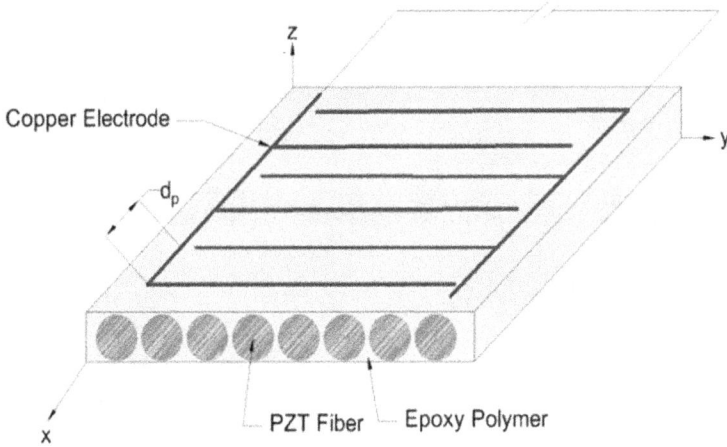

FIGURE 8.2 Diagrammatic representation of active fiber composites.

studied smart FG beams incorporated with active fiber composite as actuator material. Applications of such active fiber composites are geared towards the defense industry. This beneficial AFC finds its use in helicopter rotor blades which are normally noisy and may vibrate largely in flight due to high aerodynamic loading. These blades are embedded with AFC materials and become active rotor blades, enhancing their performance as well as that of the helicopters.

8.3 SMART FUNCTIONALLY GRADED STRUCTURES

Smart structures are structures which have self-monitoring and self-controlling capabilities. A smart structure normally consists of a passive load-bearing part such as a beam, plate, or shell which is the host structure along with an active part that may act as a sensor or actuator. These smart structures possess the ability to modify themselves in response to modified operating conditions. Piezoelectric materials and composites are utilized for active control of the smart structures. The host structure can be integrated with the smart materials without disturbing the passive stiffness characteristics of the original host structure. The smart structures having their host structures made with graded properties are conventionally known to be smart FG structures. Figure 8.3 shows a diagrammatic representation of a smart FG structure.

Highly performing smart FG structures can be designed by making the base structure of functionally graded material to satisfy multi-functional roles like increased fracture toughness, improved strength, decreased stress intensity factors, etc. These smart FG structures may be subjected to large deflections which may lead to the failure of these structures. Therefore, control of large deflections for these structures becomes necessary.

Praveen and Reddy [39] studied thermoelastic characteristics of the FG plates to ascertain the response of nonlinear stress and deflection of plates under

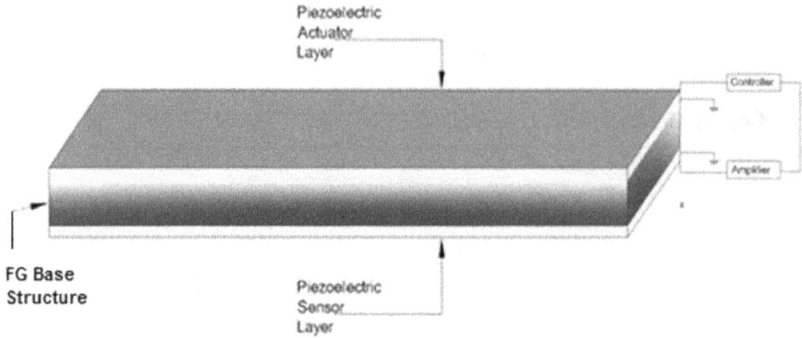

FIGURE 8.3   Diagrammatic representation of a smart FG structure.

thermo-mechanical loading conditions. Nonlinear vibration as well as dynamic analysis of FG plate in temperature environment was studied by Huang et al. [40]. Higher
order plate theory was made use of for the analysis along with the von Karman
equation considering thermal effects. Panda and Ray [41] studied the nonlinear
static behavior of smart FG plates utilizing piezoelectric fiber reinforced composite
(PFRC) as actuator and checked the performance of this actuator against the active
deformation control of functionally graded plates. Investigation was also done to see
the effects of change of fiber orientation on large deflections of FG plates. For using
the FG structures under thermal environment, Behjat et al. [42] developed an FE
model to study the static as well as dynamic characteristics of FGPM plates under
both mechanical as well as electrical loading having material property gradation
along the thickness of plate. Following first-order shear deformation theory including thermal and piezoelectric effects, results are obtained for studying the effects
of different material compositions (PZT-4H, PZT-5H) and boundary conditions for
static deformation, free vibration, and dynamic response.

8.4   ACTIVE CONSTRAINED LAYER DAMPING TREATMENT

The conventional piezoelectric materials as well as the developed piezoelectric composites (PZCs) are utilized as actuators in smart structures. For satisfactory control
of nonlinear vibrations, the control capacity of the smart materials is low which
needs to be improved. These piezoelectric materials could be utilized in the active
constrained layer damping treatment for smart control of composite structures.
This constrained layer damping (CLD) treatment (Figure 8.4) comprises a layer of
viscoelastic material between the host structure and the piezoelectric constraining
layer [43]. As the host structure experiences vibrations, the active constraining layer
prevents the viscoelastic layer from going through transverse shear deformations
and controls the transversal shear deformations, thereby promoting better damping
properties of the FG structures compared to that in passive damping. When no control voltage is applied to the constraining layer, the treatment becomes the conventional passive constrained layer damping. Also, it may be noted here that vibration

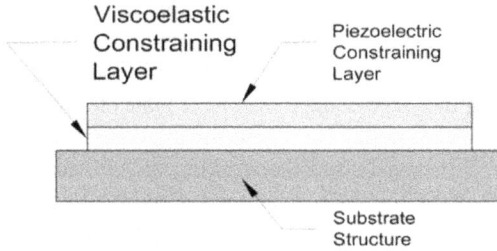

FIGURE 8.4 Constrained layer damping arrangement.

damping is achieved through shear strains developed in the damping material in the case of CLD, making it more effective compared to that in extensional damping [44].

For vibration control, the use of ACLD treatment in smart structures plays an important role and has been an attraction for investigation by researchers. A theoretical and experimental investigation was carried out by Baz and Ro [43] for computing the performance characteristics of active CLD treatment.

Baz [45] studied the damping of vibration of beams with ACLD treatment by developing a stable and global boundary control rule, and a study of the effectiveness of ACLD treatment on the vibration control of beams was carried out. Subsequently, ACLD analysis of a beam was carried out by Crassidis et al. [46]. A finite element (FE) analysis for the rods integrated fully with ACLD treatment was carried out by Baz [47] to control the wave propagation of rods. A model was developed [48] to enhance the damping behavior of viscoelastic material by attaching a constrained layer adopting the GHM method. Park and Baz [49] used classical and layer-wise laminated theories for FE analysis of ACLD. A theoretical and experimental analysis was done by Ray et al. [50] for linear vibration control of shells with ACLD treatment. Ro and Baz [51] studied control of vibration of plate structures using a self-sensing ACLD network and modal strain energy approach. Hau and Fung [52] analyzed the effect of various configurations of ACLD on the damping characteristics of flexible beams. Gao and Liao [53] compared the effectiveness of enhanced ACLD and self-sensing ACLD.

Kumar and Sharma [54] presented a theoretical and experimental study for active control of vibration in beams utilizing combined pre-compressed layer damping and ACLD. The performance of ACLD treatment on vibration control of FG beams by finite element method has been studied [55] and active fiber composites are used as a constraining layer material for the active CLD treatment. Maruani et al. [56] analyzed the vibration control of FG piezoelectric beams by the finite element method in which piezoelectric material (sensor and actuator) and functionally graded material were substituted by a single FGPM structure.

So far, limited research works are available in the literature regarding the suppression of large deformations and vibrations of FG plates utilizing the commercially available AFC material. Here first a finite element model is made for smart FG plates combined with AFC materials. Simulation models for the smart FG plates are developed using ANSYS software including the temperature-related material properties. The working of AFC material for suppressing nonlinear deflections of smart FG plates is examined. The best location for placing the AFC patches in the smart FG plates is suggested based on the efficient working of the active fiber composite for suppressing the large deflections of FG plates for different boundary conditions. Again, the AFC material is utilized for the constraining layer of smart FG plates for active CLD treatment and the working of AFC material to suppress the nonlinear vibrations of FG plates is examined. Few results are presented for suppressing nonlinear vibrations of FG plates for different boundary conditions.

8.5 FUNCTIONALLY GRADED MATERIAL PROPERTIES

Figure 8.5 represents a functionally graded plate whose material properties vary continuously and smoothly along its thickness (z) following the power-law as explained in Equation 8.1.

Here h represents the thickness of plate and k represents the non-negative power-law index or material grading index which depends upon the design requirements and the desired change of material profile along the thickness of the plate. The extreme lower and upper surfaces of the plate are considered to be rich respectively in metal and ceramic, and steel and alumina are taken as the constituents for the FG plate. The material parameters of these constituents considered here are explained in Table 8.1 [57].

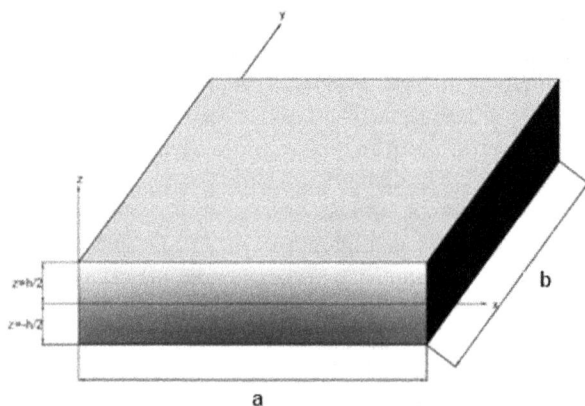

FIGURE 8.5   Schematic representation of a functionally graded plate.

TABLE 8.1
Properties of Constituent Materials for FG Plate [57]

Constituents	Young's Modulus	Density	Poisson's Ratio
Steel (SUS304)	201.04 GPa	8166 kg/m³	0.32
Alumina (Al$_2$O$_3$)	349.55 GPa	3800 kg/m³	0.26

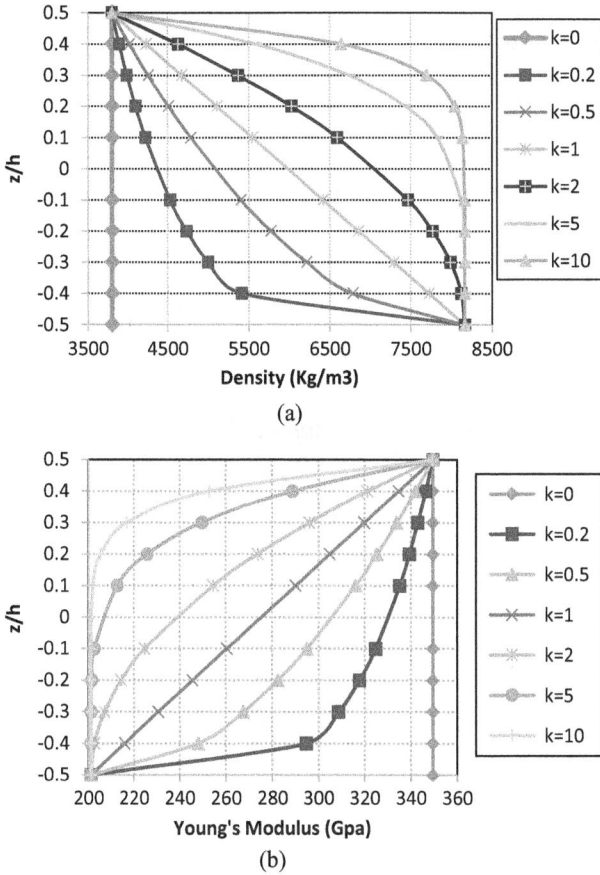

FIGURE 8.6 Variation of (a) density (b) Young's modulus along plate thickness for different power-law index (k).

The varying density (ρ) and Young's modulus (E) along the FG plate thickness are computed and are shown in Figure 8.6 for various power-law indexes. Note that the desired variation of properties for the FG plate can be obtained using specific values of the power-law index.

8.5.1 MATERIAL PARAMETERS UNDER THERMAL ENVIRONMENT

Functionally graded materials are very frequently used for applications involving elevated temperature conditions. Therefore, in this analysis, properties of the constituent materials of the FG plate are designed for elevated temperatures. For the plate, the properties of the constituent materials like density (ρ), Young's modulus (E), Poisson's ratio (υ), and thermal expansion coefficient (α), etc., are considered both temperature as well as position dependent. These material parameters are decided to

TABLE 8.2

Temperature Coefficients for Steel [57]

Parameters	P_0	P_{-1}	P_1	P_2	P_3
α (1/K)	12.33×10^{-6}	0	8.086×10^{-4}	0	0
E (Pa)	201.04×10^{9}	0	3.079×10^{-4}	-6.534×10^{-7}	0
υ	0.326	0	-2.002×10^{-4}	3.797×10^{-7}	0
ρ (kg/m³)	8166	0	0	0	0

TABLE 8.3

Temperature Coefficients for Alumina [57]

Parameters	P_0	P_{-1}	P_1	P_2	P_3
α (1/K)	6.8269×10^{-6}	0	1.838×10^{-4}	0	0
E (Pa)	349.55×10^{9}	0	-3.853×10^{-4}	4.027×10^{-7}	-1.673×10^{-10}
υ	0.26	0	0	0	0
ρ (kg/m³)	3800	0	0	0	0

change along the plate's thickness according to a modified power-law equation and temperature function as explained below [57]:

$$P = \left(P_u(T) - P_L(T)\right)\left(\frac{z}{h} + \frac{1}{2}\right)^k + P_L(T)$$

$$P = P_0\left(P_{-1}T^{-1} + 1 + P_1T + P_2T^2 + P_3T^3\right) \tag{8.4}$$

The temperature coefficients (P_0, P_{-1}, P_1, P_2, and P_3) for the constituents are described in Tables 8.2 and 8.3. These coefficients along with the properties of constituent materials as explained in Table 8.1 are used for computing the temperature- and position-dependent properties of materials for the FG plate.

8.6 ANSYS MODEL DEVELOPMENT

A functionally graded plate having dimensions "a" and "b" and combined with AFC patches of thickness "hp" on its top surface is shown in Figure 8.7.

For this smart functionally graded plate, a model for simulation is developed in an ANSYS (APDL) environment. Shell 281 (8 node) element is chosen for the discretization of the model as this element is most suited to analyze thin to moderately thick structures with lay-up options and possesses the ability to model composite as well as functionally graded structures with layered applications to perform linear as well

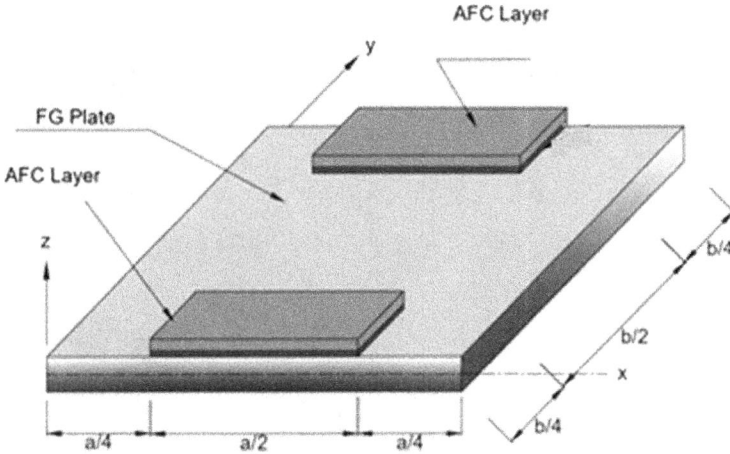

FIGURE 8.7 Representation of smart functionally graded plate combined with AFC patches.

TABLE 8.4
Convergence Analysis for Optimizing Number of Layers

| No. of Layers | Maximum Nonlinear Non-Dimensional Deflection $\left(\bar{w} = \dfrac{w}{h} \right)$ | |
	Clamped-Clamped	Simply Supported
5	0.516	0.215
10	0.562	0.223
15	0.562	0.223
20	0.562	0.223

as nonlinear analysis. This element also has the capability to accurately model layers with different material properties. The functionally graded plate considered here for analysis is made up of steel (metal) and alumina (ceramic), where the metal-rich surface is considered at the bottom while the top is ceramic rich. Material parameters of the constituents of the FG plate are mentioned in Table 8.1. Accordingly, the varying Young's modulus and density along the FG plate thickness for negligible temperature variation are computed and are as illustrated in Figure 8.6 for various power-law indexes.

The ANSYS simulation model for the smart FG plate is developed considering various layers for the substrate FG plate. For this a convergence analysis has been carried out to optimize the number of layers essential to represent the substrate FG plate. Table 8.4 describes this convergence study for clamped as well as simply supported boundary conditions for the ANSYS simulation model. It can clearly be seen from this table that ten layers can be considered for developing the model and

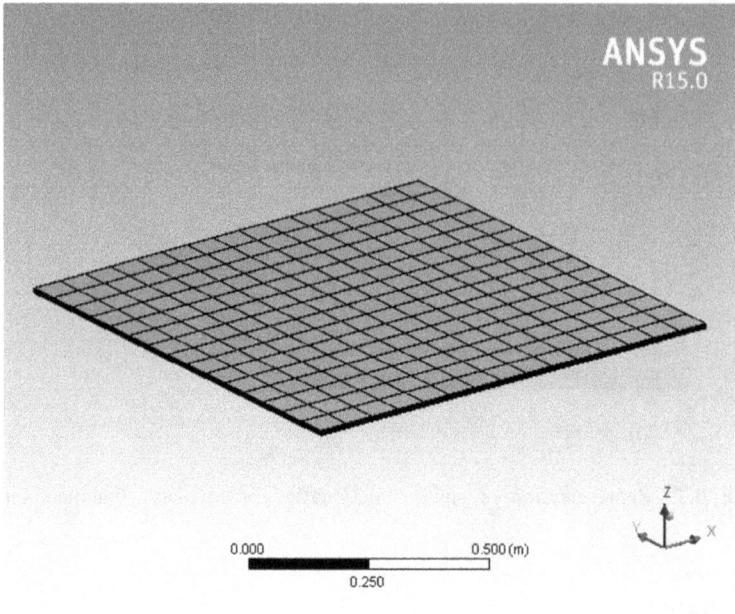

0.000   0.500 (m)

0.250

FIGURE 8.8   ANSYS simulation model for the FG plate.

henceforward a ten-layered configuration is considered for developing the models for the substrate plate. The material parameters for the various layers for the FG structure considered are calculated following the procedure as outlined earlier in Section 8.2 incorporating both the temperature and position variation along the thickness. Next, discretization of the model is carried out, and Figure 8.8 illustrates the discretized model for the FG plate using shell 281 elements. Also an optimum mesh size is decided based on the convergence study, and for the sake of brevity the details are not presented here.

8.7 MATHEMATICAL MODEL OF THE SMART FUNCTIONALLY GRADED PLATE

Figure 8.9 represents a smart FG plate combined with constrained layer patches on the top surface. The constraining and constrained layers of the CLD patches are made of AFC and a viscoelastic material respectively. The plate dimensions are a, b, and h, and h_p and h_v denote the thicknesses of the AFC and viscoelastic layers respectively. The kinematics of deformation are considered as per FSDT.

Taking θ_x, Φ_x, and γ_x as rotations of mid-plane normals of base FG plate, the viscoelastic and AFC layers, respectively, about the y-axis, θ_y, Φ_y, and γ_y are the rotations of the normals about the x-axis, displacements u, v, and w of a point in plate are

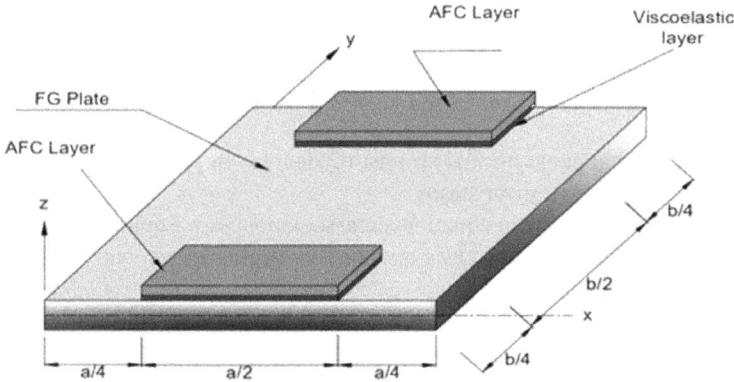

FIGURE 8.9 Functionally graded plate combined with CLD patches.

$$u(x,y,z,t) = u_0(x,y,t) + \left(z - \left\langle z - \frac{h}{2} \right\rangle \right) \theta_x(x,y,t)$$

$$+ \left(\left\langle z - \frac{h}{2} - h_v \right\rangle \right) \Phi_x(x,y,t) + \left(\left\langle z - \frac{h}{2} - h_v \right\rangle \right) \gamma_x(x,y,t)$$

$$v(x,y,z,t) = v_0(x,y,t) + \left(z - \left\langle z - \frac{h}{2} \right\rangle \right) \theta_y(x,y,t)$$

$$+ \left(\left\langle z - \frac{h}{2} - h_v \right\rangle \right) \Phi_y(x,y,t) + \left(\left\langle z - \frac{h}{2} - h_v \right\rangle \right) \gamma_y(x,y,t)$$

$$w(x,y,z,t) = w_0(x,y,t) + \left(z - \left\langle z - \frac{h}{2} \right\rangle \right) \theta_z(x,y,t)$$

$$+ \left(\left\langle z - \frac{h}{2} - h_v \right\rangle \right) \Phi_z(x,y,t) + \left(\left\langle z - \frac{h}{2} - h_v \right\rangle \right) \gamma_z(x,y,t)$$

$$(8.5)$$

in which the appropriate singularity functions are defined by the brackets, so that the displacement continuity is fulfilled between the base FG structure and the constrained layer or between the constrained layer and the AFC layer.

Displacements are arranged as a group of translation $\{d_t\}$ and rotation $\{d_r\}$ variables and are written as

$$\{d_t\} = \begin{bmatrix} u_0 & v_0 & w_0 \end{bmatrix}^T \text{ and } \{d_r\} = \begin{bmatrix} \theta_x & \theta_y & \theta_z & \Phi_x & \Phi_y & \Phi_z & \gamma_x & \gamma_y & \gamma_z \end{bmatrix}^T \quad (8.6)$$

To eliminate the shear locking problem, the selective integration rule is implemented.

In the plate, the strains at a point are indicated by vectors $\{\varepsilon_b\}$ and $\{\varepsilon_s\}$ as follows:

$$\{\varepsilon_b\} = \left[\varepsilon_x \varepsilon_y \varepsilon_{xy} \varepsilon_z\right]^T \text{ and } \{\varepsilon_s\} = \left[\varepsilon_{xz} \varepsilon_{yz}\right]^T \tag{8.7}$$

where ε_x, ε_y, and ε_z denote normal strains; ε_{xy} denotes in-plane shear strain; and ε_{xz} and ε_{yz} denote transverse shear strains.

Utilizing the Von Kármán equations in association with Equation 8.4 along with the displacement field explained by Equation 8.5, strains corresponding to a point in FG plate system are written.

To maintain consistency with the representation of strains, constitutive equations for the base FG plate are written as

$$\{\sigma_b\} = \left[C_b\right]\left(\{\varepsilon_b\} - \{\infty\}\Delta T\right), \{\sigma_s\} = \left[C_s\right]\{\varepsilon_s\} \tag{8.8}$$

and those for the PZC are

$$\{\sigma_b\} = \left[\bar{C}_b\right]\{\varepsilon_b\} - \{\bar{e}_b\}E_z \{\sigma_s\} = \left[\bar{C}_s\right]\{\varepsilon_s\} D_z = \{\bar{e}_b\}^T \{\varepsilon_b\} + \bar{\varepsilon}_{33}E_z \tag{8.9}$$

The stresses at a point in the total FG plate are written as

$$\{\sigma_b\} = \left[\sigma_x \quad \sigma_y \sigma_{xy} \quad \sigma_z\right]^T \text{ and } \{\sigma_s\} = \left[\sigma_{xz} \quad \sigma_{yz}\right]^T \tag{8.10}$$

where σ_x, σ_y, and σ_z denote normal stresses; σ_{xy} is in-plane shear stress; and σ_{xz} and σ_{yz} are transverse shear stresses.

The elastic coefficient matrices for the FG plate and PZC material are written as

$$\left[\bar{C}_b^k\right] = \begin{bmatrix} \bar{C}_{11}^k & \bar{C}_{12}^k & \bar{C}_{16}^k & \bar{C}_{13}^k \\ \bar{C}_{12}^k & \bar{C}_{22}^k & \bar{C}_{26}^k & \bar{C}_{23}^k \\ \bar{C}_{16}^k & \bar{C}_{26}^k & \bar{C}_{66}^k & \bar{C}_{36}^k \\ \bar{C}_{13}^k & \bar{C}_{23}^k & \bar{C}_{36}^k & \bar{C}_{33}^k \end{bmatrix}, \quad \left[\bar{C}_s^k\right] = \begin{bmatrix} \bar{C}_{55}^k & \bar{C}_{45}^k \\ \bar{C}_{45}^k & \bar{C}_{44}^k \end{bmatrix}, \quad \{\bar{e}_b\} = \left[\bar{e}_{31} \quad \bar{e}_{32}\bar{e}_{36} \quad \bar{e}_{33}\right]^T$$

and the piezoelectric coefficient matrices $(\{\bar{e}_b\}$ and $\{\bar{e}_s\})$ are expressed as

$$\{\bar{e}_b\} = \left[\bar{e}_{31} \quad \bar{e}_{32}\bar{e}_{36} \quad \bar{e}_{33}\right]^T \text{ and } \{\bar{e}_s\} = \left[\bar{e}_{35} \quad \bar{e}_{34}\right]^T ; k = 1, 2, 3, \ldots, N \text{ and } N+2 \tag{8.11}$$

in which the elastic co-efficient matrices for the FG plate are expressed as functions of z.

Next, the electric field is explained as $E_z = -V/h_p$ where V is voltage difference in the AFC constraining layer.

Discretization of the FG plate is done using eight noded isoparametric elements, and elemental displacement variables at a point are written as

$$\{d_t\} = \left[N_t\right]\{d_t^e\} \text{ and } \{d_r\} = \left[N_r\right]\{d_r^e\}$$

in which the vectors $\{d_t^e\}$, $\{d_r^e\}$, $[N_t]$, and $[N_r]$ are expressed as follows:

$$\{d_t^e\} = \left[\{d_{t1}^e\}^T \quad \{d_{t2}^e\}^T \cdots \quad \{d_{t8}^e\}^T\right]^T, \; \{d_r^e\} = \left[\{d_{r1}^e\}^T \quad \{d_{r2}^e\}^T \cdots \quad \{d_{r8}^e\}^T\right]^T,$$

$$\left[N_t\right] = \left[N_{t1} \quad N_{t2}\cdots \quad N_{t8}\right]^T, \; \left[N_r\right] = \left[N_{r1} \quad N_{r2}\cdots \quad N_{r8}\right]^T \qquad (8.12)$$

Strain vectors $\{\varepsilon_{bt}\}$, $\{\varepsilon_{br}\}$, $\{\varepsilon_{bnr}\}$, $\{\varepsilon_{st}\}$, and $\{\varepsilon_{sr}\}$ at a point in the element can be written as

$$\{\varepsilon_{bt}\} = \left[B_{tb}\right]\{d_t^e\}, \; \{\varepsilon_{br}\} = \left[B_{rb}\right]\{d_r^e\}, \; \{\varepsilon_{bnr}\} = \frac{1}{2}\left[B_1\right]\left[B_2\right]\{d_t^e\}$$

$$\{\varepsilon_{st}\} = \left[B_{ts}\right]\{d_t^e\}, \; \{\varepsilon_{sr}\} = \left[B_{rs}\right]\{d_r^e\} \qquad (8.13)$$

Where matrices $[B_{tb}]$, $[B_{rb}]$, $[B_{ts}]$, $[B_{rs}]$, $[B_1]$, and $[B_2]$ are the nodal strain displacement matrices.

Linear and isotropic material properties are considered for the viscoelastic layer and the stress vector $\{\sigma_s\}$ is represented by

$$\{\sigma_s\}_v = \int_0^t G(t-\tau)\frac{\partial\{\varepsilon_s\}_v}{\partial\tau}\,dt \qquad (8.14)$$

The transverse shear deflection for the viscoelastic layer is given credit to CLD for the base structure. Moreover, the extensional stiffness for the constrained viscoelastic layer becomes several orders less than that for the FG plate as well as the constraining layer. The virtual work principle applied to the FG plate is written as

$$\sum_{k=1}^{N+2}\int_{\Omega_k}\left(\delta\{\varepsilon_b^k\}^T\{\sigma_b^k\} + \delta\{\varepsilon_s^k\}^T\{\sigma_s^k\} + \rho^k\delta\{d_t\}^T\{\ddot{d}_t\}\right)d\Omega_k - \int_{\&_{N+2}}\delta E_z D_z d\&_{N+2}$$

$$-\int_A\left(\delta\{d_t\}^T\begin{bmatrix}0 & 0 & 1\end{bmatrix}^T + \delta\{d_r\}^T\left[Z_p\right]^T\right)\overline{p}\,dA = 0 \qquad (8.15)$$

Now, for any arbitrary virtual displacement the governing equations for elements combined with the ACLD patch are written as

$$\left[M^e\right]\{\ddot{d}_t^e\} + \left[K_{tt}^e\right]\{d_t^e\} + \left[K_{tr}^e\right]\{d_r^e\} + \left[K_{tsv}^e\right]\int_0^t G(t-\tau)\frac{\partial}{\partial t}\{d_t^e\}d\tau$$

$$+\left[K_{trsv}^e\right]\int_0^t G(t-\tau)\frac{\partial}{\partial t}\{d_r^e\}d\tau = \{F_t^e\} + \left(\{F_{tp}^e\} + \{F_{tpn}^e\}\right)V \qquad (8.16)$$

$$\left[K_{rt}^e\right]\{d_t^e\}+\left[K_{rr}^e\right]\{d_r^e\}+\left[K_{trsv}^e\right]^T\int_0^t G(t-\tau)\frac{\partial}{\partial t}\{d_t^e\}\,d\tau$$

$$+\left[K_{rrsv}^e\right]\int_0^t G(t-\tau)\frac{\partial}{\partial t}\{d_r^e\}\,d\tau = \{F_r^e\}+\{F_{rp}^e\}V \tag{8.17}$$

In the above equations, it is noted that the matrices $\{F_{tp}^e\}$, $\{F_{tpn}^e\}$, and $\{F_{rp}^e\}$ become null matrices when the element is not combined with ACLD treatment. The equations for the element are then mapped to get the equations of motion for the entire structure which are obtained as

$$\left[M\right]\left\{\ddot{X}_t\right\}+\left[K_{tt}\right]\{X_t\}+\left[K_{tr}\right]\{X_r\}+\left[K_{tsv}\right]\int_0^t G(t-\tau)\frac{\partial}{\partial t}\{X_t\}\,d\tau$$

$$+\left[K_{trsv}\right]\int_0^t G(t-\tau)\frac{\partial}{\partial t}\{X_r\}\,d\tau = \{F_t\}+\left(\{F_{tp}\}+\{F_{tpn}\}\right)V \tag{8.18}$$

$$\left[K_{rt}\right]\{X_t\}+\left[K_{rr}\right]\{X_r\}+\left[K_{trsv}\right]^T\int_0^t G(t-\tau)\frac{\partial}{\partial \tau}\{X_t\}\,d\tau$$

$$+\left[K_{rrsv}\right]\int_0^t G(t-\tau)\frac{\partial}{\partial \tau}\{X_r\}\,d\tau = \{F_r\}+\{F_{rp}\}V \tag{8.19}$$

in which $[M]$ is the mass matrix; $[K_{tt}]$, $[K_{tr}]$, $[K_{rt}]$, $[K_{rr}]$, $[K_{tsv}]$, $[K_{trsv}]$, and $[K_{rrsv}]$ are stiffness matrices; $\{F_{tp}\}$, $\{F_{tpn}\}$, and $\{F_{rp}\}$ are coupling vectors; $\{X_t\}$ and $\{X_r\}$ are displacement vectors, and $\{F_t\}$ and $\{F_r\}$ are mechanical force vectors globally. Finally, the time domain open loop equations for FG plates combined with ACLD patches are written as

$$\left[M^*\right]\left\{\ddot{X}\right\}+\left[C^*\right]\{\dot{X}\}+\left[K^*\right]\{X\}=\{F^*\}+\{F_p^*\}V \tag{8.20}$$

To activate the ACLD patches, a straightforward velocity feedback control rule is applied. Accordingly, the voltage requirement for each patch is written by derivatives of translational degrees of freedom which is expressed as

$$V^j = -K_d^j\dot{w} = -K_d^j\left[U^j\right]\{\dot{X}\} \tag{8.21}$$

where K_d^j is the gain of a particular jth patch and $[U^j]$ is the row vector describing location of point of sensing velocity which is backfed to the patch. Now substituting Equation 8.23 into 8.21, the closed loop governing equations for the FG plate are derived as

$$\left[M^*\right]\left\{\ddot{X}\right\}+\left[C_d^*\right]\{\dot{X}\}+\left[K^*\right]\{X\}=\{F^*\} \tag{8.22}$$

where the developed damping matrix $\left[C_d^* \right]$ is written as

$$\left[C_d^* \right] = \left[C^* \right] + \sum_{j=1}^{m} K_d^j \{ F_p^* \} \left[U^j \right]$$ (8.23)

8.8 RESULTS DISCUSSION

Here, a few results are discussed considering the models made in the previous sections. The Material properties calculated as described in Section 8.2 are used to find the results. Active fiber composite is used as the piezoelectric composite material and its effective properties considered here are [58]

C_{11} = 138.1 GPa, C_{12} = 75.15 GPa, C_{22} = 148.9 GPa, C_{66} = 39.14 GPa, e_{31} = −5.2 C/m², e_{33} = 15.1 C/m², e_{15} = 12.7 C/m². The density of the AFC material is considered as 6400 kg/m³.

8.8.1 BENDING ANALYSIS

First the models developed in the past sections are validated and for this appropriate boundary conditions are employed in ANSYS. Nonlinear center deflections are calculated for the functionally graded plate under uniform loading conditions and the deflections computed by the ANSYS simulation model are correlated with the published results [14] for various values of power index. For this comparison, the dimensions of the FG plate and the material parameters for the ANSYS simulation model are taken according to those mentioned in the literature [14]. Table 8.5 explains the comparison and verifies the present ANSYS simulation model. The maximum values of dimensionless transverse deformation for the FG plate for different power index values and boundary conditions are computed and are described in Table 8.6. It may be noticed that the deflection value increases with higher power-law index values for both clamped (CC) as well as simply supported (SS) boundary conditions.

TABLE 8.5

Center Deflection $\left(\bar{w} = \dfrac{w}{h} \right)$ for the FG Plate for Different Mesh Size and Various Power-Law Index (k) (a/h = 5)

k	Mesh Size	Ref. [14]	ANSYS Model
0.5	2 × 2	0.0357	0.0364
	4 × 4	0.0336	0.0339
1	2 × 2	0.0394	0.0402
	4 × 4	0.0373	0.0392
2	2 × 2	0.0430	0.0441
	4 × 4	0.0409	0.0415

TABLE 8.6

Maximum Transverse Deflection (w/h) of FG Substrate Plate without AFC Patches for Various Power Exponent (k)

k	0		0.2		0.5		1		2		5		10	
B.C.	CC	SS	CC	SS	CC	SS	CC	SS	CC	SS	CC	SS	CC	SS
w/h	0.59	0.21	0.62	0.23	0.64	0.24	0.67	0.25	0.69	0.26	0.71	0.27	0.73	0.28

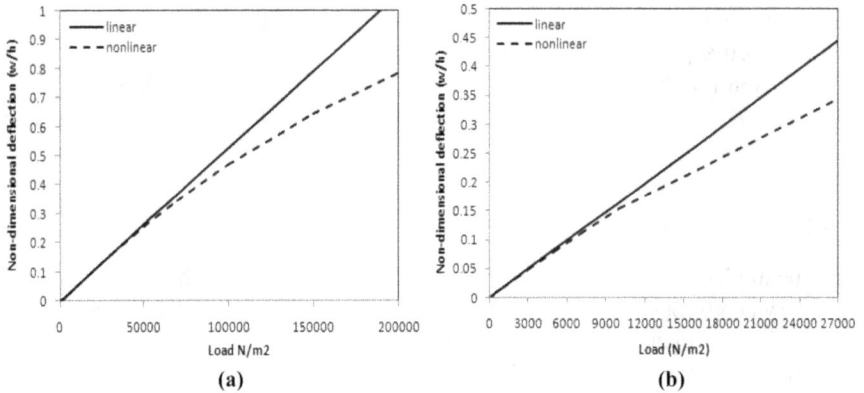

FIGURE 8.10   Load-deformation diagram of FG plate (a) clamped (b) simply supported.

The linear and nonlinear transverse deformations (w/h) for the FG plate are calculated considering clamped (CC) and simply supported (SS) boundary conditions and explained in Figure 8.10. It can be noticed from Figure 8.10a that in the case of a clamped plate, the nonlinearity is significant when the deflection value (w/h) becomes greater than 0.3 and to deal with sufficient nonlinearity, deflection (w/h) can be considered as 0.6 or higher. Similarly, for the simply supported FG plate (Figure 8.10b), the value of deflection (w/h) indicating the involvement of sufficient nonlinearity decides that a vertical pressure load 18000 N/m² may be used for the top surface of the plate for computing nonlinear deflections.

The model developed for the smart FG plate is established by matching the computed results with the results available in the literature [41]. A nonlinear study is accomplished and the dimensions and material properties of smart FG plate for the comparison may be considered according to those in the literature [41] for the FE model and ANSYS simulation model. Figure 8.11 explains the nonlinear deflection characteristics for the smart FG plates for the inactivated ($V = 0$) and activated ($V \neq 0$) AFC patches and subjected to mechanical loading. The comparison explained in Figure 8.11 indicates the agreed results, thereby validating the present methods of modeling the smart FG plates.

Analysis is performed taking the smart FG plate combined with AFC patches. The properties of the FG ingredients and AFC material considered here are as mentioned

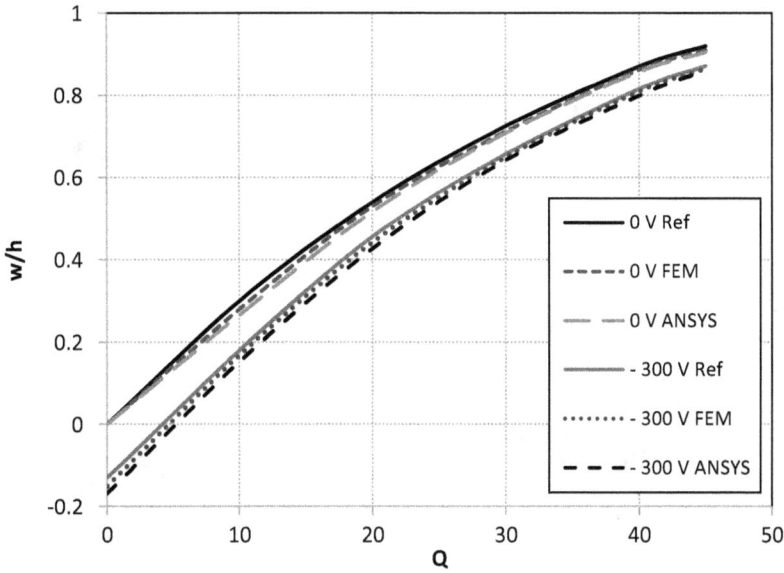

FIGURE 8.11   Central deflection for the smart FG plate ($a = b$, $k = 5$).

earlier. To examine the performance of AFC patches used as actuators for suppressing nonlinear deflections of the FG plates, results are obtained. For this purpose, FG plates combined with patches of active fiber composite material are considered in four different configurations taking the total size of AFC patches as the same in all cases. Figure 8.12 illustrates a square smart FG plate whose top surface is combined with four square AFC patches of equal size in different configurations.

When voltage is applied externally through the piezoelectric patches, the overall FG plate behaves as a smart functionally graded plate. The total length and width of AFC patches here (Figure 8.12) are taken 50% of the plate size while the thickness of the AFC patch is h_p. For computation of results, the thickness of the substrate FG plate and AFC material are considered as 10 mm and 1.5 mm respectively, and the substrate plate is taken to be of square size of side 1 m.

Transverse deflections are obtained corresponding to different power-law index values using passive ($V = 0$) and active ($V \neq 0$) AFC patches.

The non-dimensional parameters used here for displaying the results are as follows:

$$Q = \frac{Ps^4}{E_U} \text{ and } \dot{w} = \frac{w}{h}$$

E_U is the modulus of elasticity for the uppermost surface of the substrate FG plate. Simply supported and clamped boundary conditions are adopted to obtain the results.

The performance of AFC patches in the reduction of nonlinear deformation of FG plates is studied for all four different cases. For this, load-deflection curves are

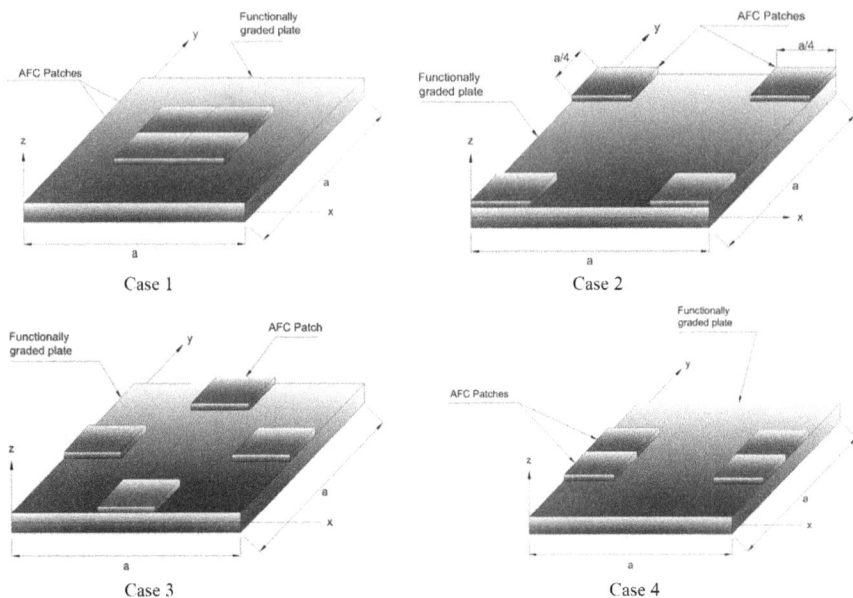

FIGURE 8.12   Various configurations of smart FG plates combined with AFC patches.

obtained in nonlinear range for the smart FG plate having AFC patches in four different configurations as described in Figure 8.12. It is noticed that when the patches are placed according to case 3 of Figure 8.12, the performance of the active AFC patches seems to be maximum in the control of nonlinear deformations of the FG plates.

Taking this best position for placing the AFC patches on the top surface of the FG plate, the performance of the patches for suppressing nonlinear deflections of the FG plates is studied for various control voltages for the FG plates (Figure 8.13).

The functionally graded plate is also modeled for thermal environment following the procedure explained in Section 8.2. The smart FG plate is modeled for a uniform temperature $T = 500$ K in which $T = T_0 + \Delta T$, T_0 being the initial temperature and ΔT being the uniform increase of temperature. Material properties used in the development of this model for thermal environment are also discussed in Section 8.2. The plate is loaded with a pressure load on its top. The maximum transverse deflections for the FG plate are obtained for various power-law indexes for the plate subjected to both temperature and mechanical loading. For an increase of power-law index values, increased deflection is observed, and by correlating these results with those presented in Table 8.6, it is noticed that the deformation of the FG plate is considerably higher for applied thermo-mechanical loading than that obtained when the plate is subjected to mechanical load only. This indicates softening behavior of the plates in elevated temperature conditions.

The performances of AFC patches when placed in the best location (case 3 of Figure 8.12) of the FG plate for various values of power index in elevated temperature

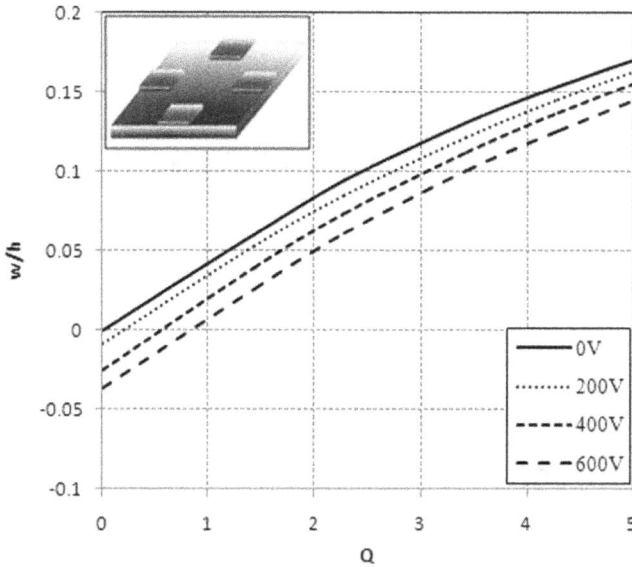

FIGURE 8.13   Center deflections of smart simply supported FG plate with applied mechanical load (Q) for various control voltages.

conditions are investigated (Figure 8.14). The voltages used for the AFC patches are indicated in the figure, and it can be observed that the AFC patches largely contribute towards deflection control of the FG plates.

8.8.2   VIBRATION ANALYSIS

The simulation model developed in ANSYS software for the smart FG plate is analyzed considering negligibly thick patches. First the natural frequencies are calculated using the ANSYS simulation model and compared with available results [14]. These results are presented in Table 8.7 and for this comparison, the geometry and constituent properties for the FG plate are considered according to those considered in the reference [14]. It is noticed that the results excellently match each other.

Natural frequencies are obtained for the smart FG plates, using the simulation model for different boundary conditions as well as power indexes with negligible thicknesses of the CLD patches. These are presented in Table 8.8. Material properties considered here for the plate analysis are explained in Section 8.2. The natural frequencies for the FG plates are written in dimensionless form using the relation

$\lambda = \omega L^2 \sqrt{\dfrac{\rho_l}{E_l} \dfrac{A}{I}}$. Simply supported boundary conditions are adopted to obtain the results.

Figure 8.15 explains the mode shapes for the clamped and simply supported smart FG plates. The nonlinear transient vibrations for the FG plates are obtained considering the sinusoidal loading and are presented in Figure 8.16.

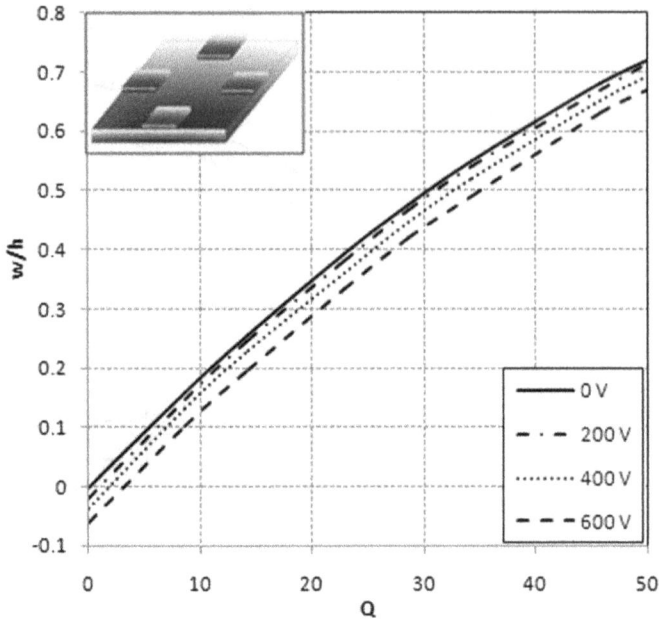

FIGURE 8.14 Center deflection of smart FG plate subjected to thermo-mechanical loading (CC).

TABLE 8.7

Frequency Parameter for Different Volume Fraction Index for Square FG Plates

N	Mode	Ref. [14]	Present Model
0.5	1	1.7566	1.7552
	2	4.2631	4.2629
1	1	1.6853	1.6732
	2	4.0609	4.0596

The smart functionally graded plate is considered with its top surface combined with patches of CLD treatment. The results are then computed for nonlinear dynamic responses of smart FG plates. In this analysis the voltage used is proportional (negative) to the velocity of a point on the FG plate analogous to the free length midpoint of patch. The control gain (K_d) is so selected as to suitably control the nonlinear vibrations. The transient responses (nonlinear) of a functionally graded plate (CC) combined with the CLD patches are presented in Figure 8.17 for passive $(K_d = 0)$ as well as active $(K_d \neq 0)$ conditions of the patches.

The patches in active condition provide satisfactory control of the nonlinear vibrations thus enhancing the damping characteristics of the FG plate. Control

TABLE 8.8

Fundamental Frequency for FG Plates

Boundary Conditions	Frequency	$k = 0.2$	$k = 1$	$k = 5$
SS1	λ_1	32.67	27.88	24.45
	λ_2	81.44	69.13	60.69

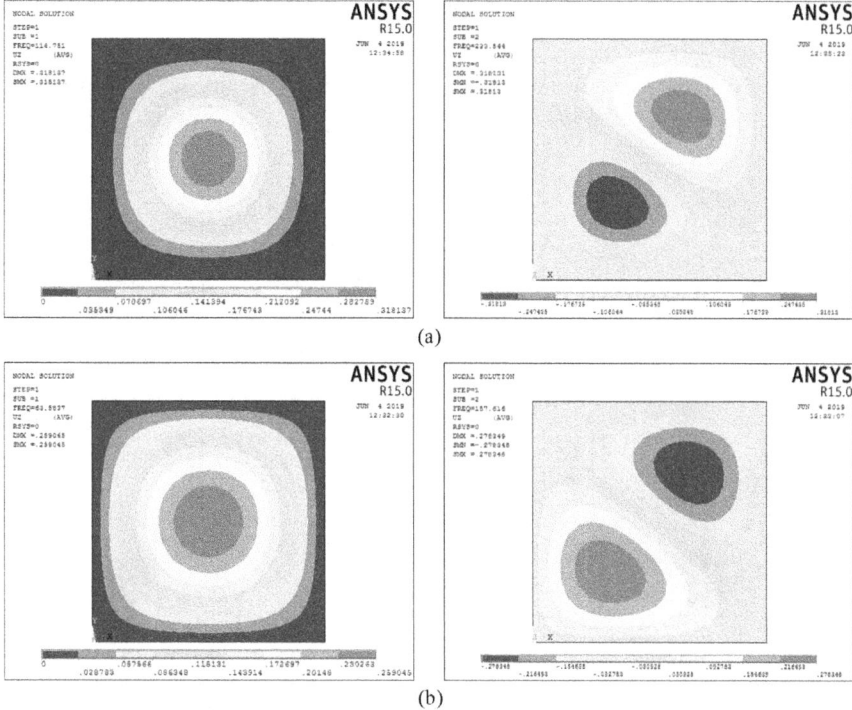

(a)

(b)

FIGURE 8.15 First and second mode shapes for (a) clamped-clamped and (b) simply supported smart functionally graded plate ($k = 1$).

voltage requirements corresponding to the gain are reasonable and are presented in Figure 8.18. The voltage applied to the patch corresponds to velocity and thus the applied voltage illustrated denotes decayed velocity of a point on the FG plate with time. The phase plot diagram for the clamped FG plate is shown in Figure 8.19, and this indicates the stability of the overall functionally graded structure.

Similarly, results are also obtained for studying the transient behavior for FG plates combined with CLD patches for simply supported boundary conditions. The transient responses, control voltage requirements, and the phase plot diagram of the simply supported FG plate combined with the patches are presented in Figure 8.20. It

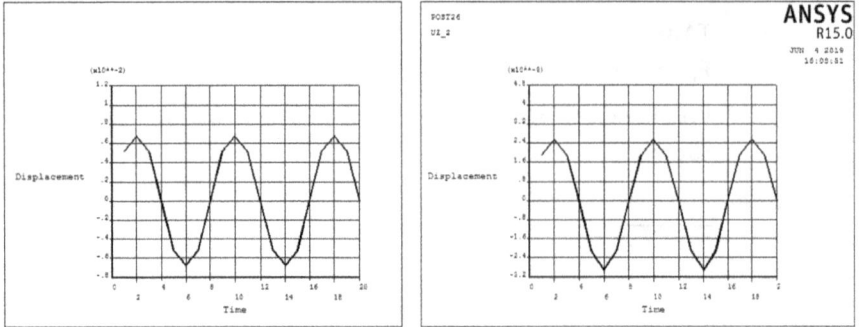

FIGURE 8.16 Transient response of (a) clamped-clamped and (b) simply supported FG plate with negligibly thick ACLD patches.

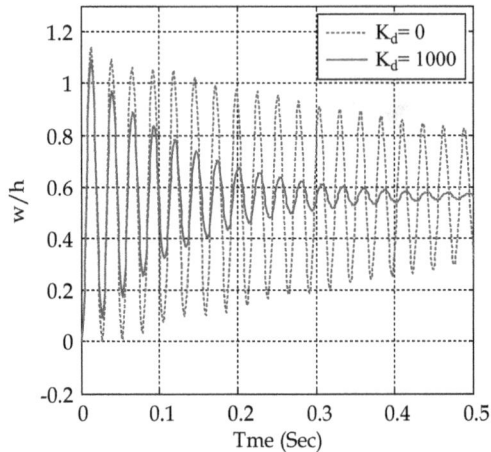

FIGURE 8.17 Transient responses of clamped functionally graded plate (active and passive).

may be observed that the patches provide satisfactory control of the nonlinear vibrations and thus enhance the damping characteristics of the FG plate.

8.9 CONCLUSIONS

Smart FG plate analysis has been discussed in this chapter. Finite element models are developed for smart functionally graded plates considering the thermal environment. First, the performance of AFC material for suppressing the nonlinear deformations of the FG plates is investigated. The AFC material is used as an actuator layer for the smart FG plates. Von Kármán nonlinear strain displacement equations are used along with FSDT for developing a model for the structure, and simulation models are developed using ANSYS software for the smart FG plates. Different boundary conditions and power-law indexes are used for evaluations of numerical results. The computed results indicate that the nonlinear deflections of

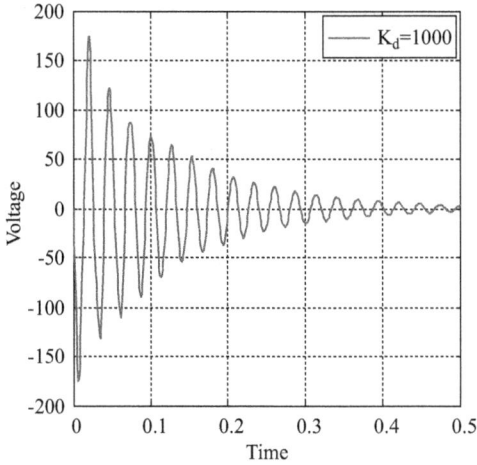

FIGURE 8.18   Voltage requirements for active CLD patch for transient vibrations of FG substrate plate (clamped).

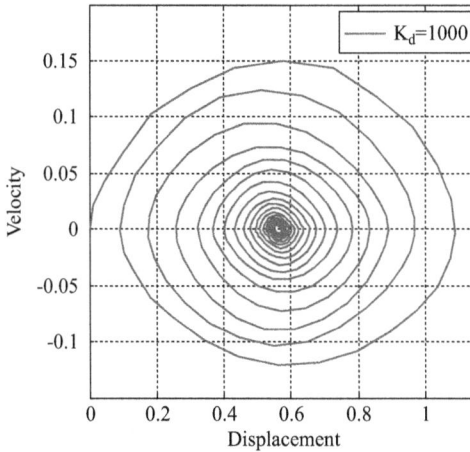

FIGURE 8.19   Phase plot diagram of clamped FG substrate plate undergoing active constrained layer damping.

the smart FG plates are higher for the case of thermo-mechanical loading when compared to only mechanical loading for both clamped as well as simply supported boundary conditions and also demonstrate softening behavior of the plates in elevated temperature conditions. It is observed that the active AFC patches perform satisfactorily to control the nonlinear deformations of these smart FG plates. The optimum placement of AFC patches is obtained for getting the best control of deformation of the FG plates in nonlinear range for both clamped and simply supported boundary conditions. Next, a nonlinear vibration damping study for the

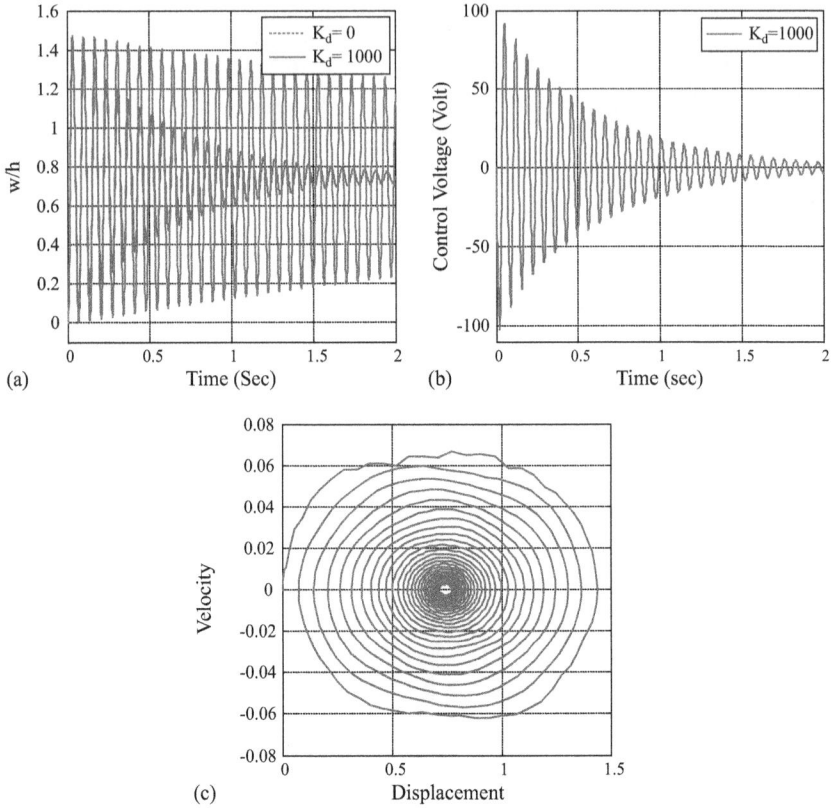

(a)   (b)   (c)

FIGURE 8.20 (a) Transient responses (b) voltage requirements (c) phase plot diagram for smart simply supported FG plate.

smart FG plates has been presented. An FE model is developed for the smart FG plates combined with CLD patches. For time domain analysis, the GHM technique is applied for modeling the viscoelastic layer of the CLD patches. For active damping, the velocity feedback control algorithm is used. Simulation models developed in ANSYS software for the smart FG plates with negligibly thick CLD patches are used to validate the defined FE model. Transient response results for the FG plates explain that the active CLD treatment undoubtedly enhances the damping behavior of FG plates as compared to passive damping for controlling their nonlinear vibrations.

REFERENCES

1. Pindera, M. J., Arnold, S. M., Aboudi, J., Hui. D. Use of composites in functionally graded material. *Composites Engineering* 1994. 4: 1–145.
2. Koizumi, M. The concept of FGM. *Ceramic Transactions Functionally Graded Materials.* 1993. 34: 3–10.

3. Sata, N. Characteristic of SiC-TiB2 composites as the surface layer of SiC-TiB2-Cu Functionally gradient material produced by self-propagating high temperature synthesis. *Ceramic Transactions Functionally Graded Materials*, 1993. 34: 109–116.

4. Yamaoka, H., Yuki, K., Tahara, K., Irisawa, T., Watanabe, R., Kawasaki, A. Fabrication of functionally gradient material by slurry stacking and sintering process. *Ceramic Transactions Functionally Gradient Materials*, 1993. 34: 165–172.

5. Rabin, B. H., Heaps, R. J. Powder processing of Ni-Al2O3 FGM. *Ceramic Transactions Functionally Gradient Materials*, 1993. 34: 173–180.

6. Kieback, B., Neubrand, A., Riedel, H. Processing techniques for functionally graded materials. *Materials Science and Engineering*, 2003. 362: 81–105.

7. Birman, V. Stability of functionally graded shape memory alloy sandwich panels. *Smart Materials and Structures*, 1997. 6: 278–286.

8. FGM Forum. *Survey for Application of FGM*, Society of Non-Tradition Technology, Tokyo, Japan, 1991.

9. Wakashima, K., Hirani, T. and Niino, M. Space applications of advanced structural materials. *ESA Secial Publication*, 1990: 97–102.

10. Kim, J. H., Paulino, G. H., Finite element evaluation of mixed mode stress intensity factors in functionally graded materials. *International Journal for Numerical Methods in Engineering*, 2002. 53(8): 1903–1935.

11. Chung, Y. L., Chen, W. T. Bending behavior of FGM-coated and FGM-undercoated plates with two simply supported opposite edges and two free edges. *Composite Structures*, 2007. 81: 157–167.

12. Li, X. F., Wang, B. L., Han, J. C. A higher-order theory for static and dynamic analyses of functionally graded beams. *Archive of Applied Mechanics*, 2010. 80: 1197–1212.

13. Reddy, J. N., Chin, C. D. Thermoelastical analysis of functionally graded cylinders and plates. *Journal of Thermal Stresses*, 1998. 21(6): 593–626.

14. Talha, M., Singh, B. N. Static response and free vibration analysis of FGM plates using higher order shear deformation theory. *Applied Mathematical Modelling*, 2010. 34: 3991–4011.

15. Alshorbagy, A. E., Eltaher, M. A., Mahmoud, F. F. Free vibration characteristics of a functionally graded beam by finite element method. *Applied Mathematical Modelling*, 2011. 35: 412–425.

16. Mehta, R., Balaji, P. S. Static and dynamic analysis of functionally graded beam. *Indian Journal of Research*, 2013. 2: 80–85.

17. Xiang, H. J., Yang, J. Free and forced vibration of a laminated FGM Timoshenko beam of variable thickness under heat conduction. *Composites: Part B*, 2008. 39: 292–303.

18. Reddy, B. S., Kumar, J. S., Reddy, C. E., Reddy, K. V. K. Static analysis of functionally graded plates using higher-order shear deformation theory. *International Journal of Applied Science and Engineering*, 2014. 12: 23–41.

19. Ferreira, A. J. M., Batra, R. C., Roque, C. M. C., Quin, L. F., Martins, P. A. L. S. Static analysis of functionally graded plates using third order shear deformation theory and a meshless method. *Composite Structures*, 2005. 69: 449–457.

20. Jing, L., Ming, P. J., Zhang, W. P., Fu, Li., Cao, Y. Static and free vibration analysis of functionally graded beams by combination Timoshenko theory and finite volume method. *Composite Structures*, 2016. 138: 192–213.

21. Pydah, A., Sabale, A. Static analysis of bi-directional functionally graded curved beams. *Composite Structures*, 2017. 160: 867–876.

22. Bajpai, R. B. Applications of smart materials in present day engineering. *Journal of Department of Applied Sciences and Humanities*, 2011. XI: 8–10.

23. Ray, M. C., Mallik, N. Performance of smart damping treatment using piezoelectric fiber reinforced composites. *AIAA Journals*, 2005. 45: 184–193.

24. Gentilman, R., McNeal, K., Schmidt, G., Piezzochero, A., Rossetti George, A. Enhanced performance active fiber composites. *Smart Materials and Structures*, 2003. 5054: 350–359.

25. Lin, X. J., Zhou, K. C., Zhang, X. Y., Zhang, D. Development, modeling and application of piezoelectric fiber composites. *Transactions of Nonferrous Metals Society of China*, 2013. 23: 98–107.

26. Mallik, N., Ray, M. C. Effective coefficients of piezoelectric fiber reinforced composites. *AIAA Journal*, 2003. 41: 704–710.

27. Ray, M. C., Pradhan, A. K. The performance of vertically reinforced 1–3 piezoelectric composites in active control of smart structures. *Smart Materials and Structures*, 2006. 15: 631–641.

28. Li, J. Y., Dunn, M. L. Micromechanics of Magneto eletroelastic Composite material: Average fields and effective behavior. *Journal of Intelligent Material Systems and Structures*, 1998. 9: 404–416.

29. Sakthivel, M., Arockiarajan, A. Studies on effective fiber and matrix poling characteristics of 1–3 piezoelectric composites. *International Journal of Structural Changes in Solids*, 2011. 3: 1–19.

30. Medeiros, R., Moreno, M. E., Marques, F. D., Tita, V. Effective properties evaluation for smart composite materials. *Journal of the Brazilian Society of Mechanical Sciences and Engineering*, 2012. 34: 362–370.

31. Tawakol, A. E. Evaluation of the effective electromechanical properties of unidirectional piezocomposites using different representative volume elements. *International Journal of Mechanical and Mechatronics Engineering*, 2015. 15: 21–29.

32. Chan, H. L. W., Unsworth, J. Simple Model for piezoelectric ceramic/polymer 1–3 composites used in ultrasonic transducer applications. *IEEE Transactions on Ultrasonics, Ferroelectrics and Frequency Control*, 1989. 36: 434–441.

33. Smith, W. A., Auld, B. A. Modelling 1–3 composites piezoelectrics: Thickness mode oscillations. *IEEE Transactions on Ultrasonics, Ferroelectrics and Frequency Control*, 1991. 38: 40–47.

34. Ray, M. C., Pradhan, A. K. On the use of vertically reinforced 1–3 piezoelectric composites for hybrid damping of laminated composite plates. *Mechanics of Advanced Materials and Structures*, 2007. 14: 245–261.

35. Bent, A. A., Hagood, N. W. Piezoelectric fiber composites with integrated electrodes. *Journal of Intelligent Material Systems and Structures*, 1997. 8: 903–919.

36. Kar-Gupta, R., Venkatesh, T. A. Electromechanical response of 1–3 piezoelectric composites: An analytical model. *Acta Materialia*, 2007. 55: 1093–1108.

37. Ivanov, I. V. Finite element analysis and modeling of Active fiber composites including damages, *Computational Materials Science*, 2011. 50: 1276–1282.

38. Tripathy, A., Panda, R. K., Das, S., Sarangi, S. K. Static analysis of smart functionally graded beams. *Spvryan's International Journal of Engineering Sciences and Technology*, 2015. 2(4): 1–6.

39. Praveen, G. N., Reddy, J. N. Nonlinear transient thermoelastic analysis of functionally graded ceramic-metal plates. *International Journal of Solids Structures*, 1998. 35 (33): 4457–4476.

40. Huang, X. L., Shen, H. S. Nonlinear vibration and dynamic response of FGP in thermal environment. *International Journal of Solids and Structures*, 2004. 41: 2403–2427.

41. Panda, S., Ray, M. C. Nonlinear analysis of smart functionally graded plates integrated with a layer of piezoelectric fiber reinforced composite. *Smart Materials and Structures*, 2006. 15(6): 1595–1604.

42. Behjat, B., Salehi, M., Armin, A., Sadighi, M., Abbasi, M. Static and dynamic analysis of functionally graded piezoelectric plates under mechanical and electrical loading. *Scientia Iranica*, 2011. 18(4): 986–994.

43. Baz, A., Ro, J. Optimum design and control of active constrained layer damping. *ASME Journal of Vibrations and Acoustics*, 1995. 117B: 135–144.

44. Sarangi, S. K., Ray, M. C. Smart damping of geometrically nonlinear vibrations of laminated composite beams using vertically reinforced 1–3 piezoelectric composites. *Smart Materials and Structures*, 2010. 19: 075020.

45. Baz, A. Dynamic boundary control of beams using active constrained layer damping. *Mechanical Systems and Signal Processing*, 1997. 11: 811–825.

46. Crassidis, J. L., Baz, A., Wereley, N. Control of active constrained layer damping", *Journal of Vibration and Control*, 2000. 6: 113–136.

47. Baz, A. Spectral finite element modeling of the longitudinal wave propagation in rods treated with active constrained layer damping. *Smart Materials and Structures*, 2000. 9: 372–377.

48. Lam, M. J., Inman, D. J., Saunders, W. R. Hybrid damping models using the Golla-Hughes-McTavish method with internally balanced model reduction and output feedback. *Smart Materials and Structures*, 2000. 9: 362–371.

49. Park, C. H., Baz, A. Comparison between Finite Element Formulations of Active Constrained Layer Damping using Classical and Layer wise Laminate Theory. *Finite Elements in Analysis and Design*, 2001. 37: 35–56.

50. Ray, M. C., Oh, J., Baz, A. Active Constrained Layer Damping of Thin Cylindrical shells. *Journal of Sound and Vibration*, 2001. 240: 921–935.

51. Ro, J., Baz, A. Vibration control of plates Using Self-sensing Active Constrained layer damping networks. *Journal of Vibration and Control*, 2002. 8: 833–845.

52. Hau, L. C., Fung, E. H. K. Effect of ACLD treatment configuration on damping performance of a flexible beam. *Journal of Sound and Vibration*, 2004. 269: 549–567.

53. Gao, J. X., Liao, W. H. Vibration analysis of simply supported beams with enhanced self-sensing active constrained layer damping treatments. *Journal of Sound and Vibration*, 2005. 280: 329–357.

54. Kumar, S., Sharma, R. K., Active vibration control of beams by combining precompressed layer damping and ACLD treatment: Theory and Experimental implementation. *Journal of Vibrations and Acoustics*, 2013. 133(6): 4005028.

55. Panda, R. K., Nayak, B., Sarangi, S. K. Active vibration control of smart functionally graded beams. *Procedia Engineering*, 2016. 144: 551–559.

56. Maruani, J., Bruant, I., Pablo, F., Galimard, L. A numerical efficiency study on the active vibration control for a FGPM beam. *Composite Structures*, 2018. 182: 478–486.

57. Trinh, L. C., Vo, T. P., Thai, H., Nguyen, T. K. An analytical method for the vibration and buckling of functionally graded beams under mechanical and thermal loads. *Composites Part-B*, 2016. 100: 152–163.

58. Sarangi, S. K., Basa, B. Nonlinear finite element analysis of smart laminated composite sandwich plates. *International Journal of Structural Stability and Dynamics*, 2014. 14: 1350075.

9 Dynamic Analysis of a Porous Sandwich Functionally Graded Material Plate with Geometric Nonlinearity

Simran Jeet Singh and Suraj Prakash Harsha

CONTENTS

DOI: 10.1201/9781003097976-9

9.1   INTRODUCTION

Functionally graded material (FGM) is a passive smart material in which the properties of two materials are graded in one particular direction. The idea of FGM was substantially advanced in the early 1980s in Japan during a spaceplane project [1]. Since then, worldwide research efforts have been initiated by FGM principles and are applied to metals, ceramic production, and natural composites to create improved segments with unrivaled actual properties [2]. The variation in material properties is then accomplished by different approaches depending on the applications and working environment. There are various techniques of production; the most common are plasma vapor deposition and powder metallurgy for producing thin and bulk FGM, respectively.

Further, sandwich structures, due to low weight and high strength, found applications in aerospace, automobiles, defense, etc. Functionally graded sandwich plates are however one of the focused areas of research for various researchers. This is because the separation of face sheets that occurs as a result of working in harsh conditions in the conventional sandwich plate can be avoided by making use of functionally graded face sheets. This is due to the continuous variation of constituent materials at the interfaces in sandwich FGM plates. These sandwich FGMs, in addition to separation, also relieve stresses at the interfaces and hence avoid stress concentration. The analysis of sandwich FGM plates by various researchers is focused on static and dynamic analysis under environmental factors affecting the plate behavior.

The static analysis of the sandwich FGM plate had been done to predict the flexural behavior viz. deflections and stresses. In this context, Zenkour [3–5] proposed theories for determining the exact solution for perfect and porous sandwich FGM plates. Tounsi et al. [6] performed thermo-elastic bending analysis of sandwich FGM plates using refined trigonometric shear deformation theory (TSDT). Taibi et al. [7] proposed a simple refined shear deformation for the bending response of a sandwich FGM plate resting on a two-parameter foundation. Bessaim et al. [8] did the bending analysis by proposing a new higher-order and normal shear deformation theory that takes into account the thickness stretching effect. Sciuva and Sorrenti [9, 10] proposed a refined zigzag theory for the analysis of sandwich and laminated plates. Additionally, static analysis is performed to study the effect of geometric parameters and environmental effects. In this context, a sandwich FGM plate subjected to uniformly distributed load was studied by Demirhan and Taskin [11]. Further, implementing the state-space approach, bending and free vibration analysis was investigated by Demirhan et al. [12] for Levy's FG porous plate. Sobhy [13] analyzed FGM plates supported with different boundary conditions and resting on elastic foundations in a thermal environment. The temperature-dependent material properties of the plate were assumed to be graded according to a power-law distribution. Singh and Harsha [14] studied the porous sandwich FGM plate subjected to thermo-mechanical loading in order to predict the flexural and stress responses. Similarly, a lot of research was done on the sandwich FGM plate to assess the flexural behavior [15].

Further, as per dynamic analysis, primarily the research is focused on determining the inherent characteristics of the sandwich plate viz. the fundamental frequency through free vibration analysis. In this context, Zenkour [16] determined the fundamental frequency of the sandwich plate by considering various geometrical parameters. Kurpa and Shmatko [17] incorporated the Ritz method and the R-functions theory to calculate the fundamental frequency of FG sandwich plates and shells. Hadji et al. [18] obtained a closed-form solution for free vibration analysis of sandwich plate using Navier's method. The Galerkin–Vlasov method is employed to determine the frequency for the sandwich S-FGM [19] and E- FGM [20, 21] plate resting on the Pasternak foundation. Further, Singh and Harsha [22] studied the effect of porosity on free vibration responses for sandwich S-FGM plates. The problem [19–22] was formulated using the variational principle, and the effect of different types of boundary conditions was studied. Wattanasakulpong et al. [23] implemented the Ritz method to investigate free vibration responses in a thermal environment. Meiche et al. [24] studied sandwich plate using the new proposed hyperbolic theory and used Navier's method to obtain results. On the same track, Thai et al. and Nguyen et al. presented a new first-order [25] and inverse trigonometric [26] shear deformation theory. The results obtained from the new theories [24–26] were in good harmony with the previous results. Xiang et al. [27] demonstrated the accuracy and efficiency of the meshless global collocation method for the analysis of sandwich plates. Thus, the literature on free vibration analysis of a sandwich plate is available either on newly proposed theories [8, 24–26, 28, 29] or on new methods of formulating and solving [16, 27, 30, 31] the free vibration problem.

Furthermore, the free vibration analysis assists in predicting the forced vibration response of the sandwich structures under resonance conditions. The nonlinear forced vibration response can be seen as a real-time response for any system when subjected to harmonic excitation with real-time environmental conditions. Although a lot of literature on the dynamic analysis of isotropic rectangular plates [32–35] is available, on FGMs, the available literature is scarce. In this framework, the well-known books by Amabili [36] and Duc [37] provide a piece of in-depth knowledge on the nonlinear vibration of plates and shells. Amabili [38–40] and his co-workers employed a multimodal energy approach to examine the time and frequency responses. The AUTO pseudo-arc-length continuation and collocation technique were used for bifurcation analysis, whereas Duc [41, 42] and his co-workers implemented the airy stress function in conjunction with the Galerkin method in most of their research to derive governing nonlinear coupled ODEs. Then, a numerical integration scheme is used to analyze the nonlinear dynamic responses.

Alijani et al. [43] examined a simply supported FGM plate with movable edges in a thermal environment. To classify and detect bifurcations and complex nonlinear dynamics, Lyapunov exponents were obtained. Yang and Shen [44] analyzed the initially stressed FGM plates for free and forced vibration analysis. The differential quadrature method (DQM) was employed to obtain a solution. Huang and Shen [45] analyzed the nonlinear dynamic response and frequencies for FGM plates in the thermal environment using an improved perturbation technique. Praveen and Reddy [46] studied the effect of boundary conditions and other geometric parameters on the non-dimensional time responses for different combinations of materials in FGM. The plate was assumed to be subjected to suddenly applied uniform loading. Upadhyay and Shukla [47] presented a nonlinear vibration analysis for FGM skew plates subjected to different types of loadings. Hao et al. [48] analyzed nonlinear dynamic responses for an FGM rectangular plate. The plate was acted upon by the transverse and in-plane harmonic excitations. It was found that the periodicity of the plate is largely affected by the time-dependent temperature. Akbarzadeh et al. [49] analyzed the combined effect of thermal and mechanical load on an FGM rectangular plate. The analysis was performed using the hybrid Fourier Laplace transform method. Duc and Cong [50] analyzed the time responses and frequency–amplitude relations using a semi-analytical technique, viz. the Galerkin method with Airy's stress function. They studied the effect of considering inertia forces due to rotations and concluded that the effect is negligible when performing nonlinear dynamic analysis. Jung et al. [51] studied the linear free and forced vibration analysis of a sigmoid FGM (S-FGM) plate resting on a Pasternak foundation and solved equations using Navier's method. Insofar as sigmoid function-based sandwich FGM plates are concerned, recently, nonlinear dynamic analysis [52, 53] was performed on the plate. On the same track, they performed analysis in a thermal environment [54]. In addition, large-amplitude forced vibration of porous S-FGM plates with porosity effect was studied by Wang and Zu [55] in which even and uneven porosity distributions were considered.

From the above discussed literature, it is established that no work has been reported on studies of the thermal and porosity effect on nonlinear dynamic responses for sandwich FGM plates. This study is certainly vital to analyze the inherent characteristics of sandwich FGM plates under the effects of design parameters. The present chapter, in a thermal environment, analyzed the dynamic characteristics of porous sandwich S-FGM plates considering geometric nonlinearity. This chapter is also new in its assessment of the plate behavior in terms of predicting the periodicity such as periodic, quasi-periodic, chaotic, etc., and analyzing the nonlinear frequency–amplitude relation. The aforementioned prediction aids researchers in the active control of plates and shells under dynamic loading.

The chapter is prepared as follows: in Section 9.1, a general introduction to the sandwich FGM plates and available literature is discussed. In Section 9.2, porosity and temperature distribution are described. In Section 9.3, the material properties of sandwich FGM plates along with the thermo-mechanical constitutive relation are defined. The kinematic equations and governing differential equations are presented in Section 9.4. Then, a semi-analytical solution using the Galerkin method is demonstrated in Section 9.5. In Section 9.6, governing equations for nonlinear free and forced vibration analysis are presented. In Sections 9.7 and 9.8, validation, convergence, and a parametric study are exhibited. Finally, concluding remarks are provided in Section 9.9. References are then provided.

9.2 POROSITY AND TEMPERATURE DISTRIBUTION

9.2.1 Porosity Models

The prevalence of pores at some stage in the fabrication procedure is modeled via different mathematical functions [56–59] which include transcendental functions, exponential functions, polynomial functions, etc. As depicted in Figure 9.1b and Figure 9.1c, homogenous and symmetric but uneven distribution is usually employed by various researchers in the analysis. But, in the present study, a new porosity distribution is proposed which will capture the realistic variation of the pores across the thickness of the plate as depicted in Figure 9.1d. Various porosity variations implemented in this study are:

- **Homogenous/even distribution (P – 1)**: the micro-voids are distributed consistently in the thickness direction which will severely decrease the overall stiffness and density of the plate as shown in Figure 9.1b. The mathematical function for $P-1$ porosity distribution is represented as follows:

$$P-1 = \frac{e}{2} \tag{9.1}$$

- **Uneven symmetric distribution (P – 2)**: the micro-voids are distributed according to the mode function in the thickness direction which will gradually reduce the overall stiffness and density of the plate as shown in Figure 9.1c. The mathematical function for $P-2$ porosity distribution is represented as follows:

$$P-2 = \frac{e}{2}\left(1 - \frac{2\,|z|}{\hbar}\right) \tag{9.2}$$

- **Uneven non-symmetric distribution (P – 3)**: the micro-voids are distributed consistently according to sigmoid function in the thickness direction which will smoothly and non-uniformly decrease the overall stiffness and

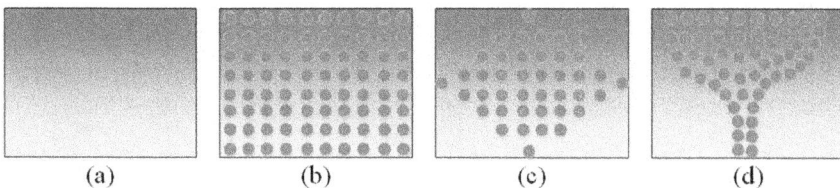

FIGURE 9.1 Porosity distribution across the sandwich FGM plate (a) perfect, (b) P – 1, (c) P – 2, (d) P – 3.

density of the plate as shown in Figure 9.1d. However, this proposed model will be one of the most realistic porosity distribution models that can be proved to be more accurate. The mathematical function for $P - 3$ porosity distribution is represented as follows [14, 22]:

$$p - 3 = \frac{e}{2} \left[\frac{1}{1 + \exp\left[\frac{-pz}{\hbar} \right]} \right]$$

(9.3)

Where e is a small fraction representing the voids present in a plate, p is a number (≥ 0) representing the volume fraction exponent of constituent material, and $\hbar$ is the thickness of a plate.

9.2.2 TEMPERATURE DISTRIBUTION

The temperature typically varies according to some mathematical concepts in the transverse direction or is regulated by certain scientific principles. In the present study, two different types of temperature distribution are assumed:

9.2.2.1 Uniform Temperature Distribution
In this case, the constant temperature is assumed to be distributed across the thickness of the plate. Thus, at a certain thickness, the following mathematical function is implemented to represent uniform temperature distribution:

$$T(z) = T_0 + \Delta T$$

(9.4)

9.2.2.2 Nonlinear Temperature Distribution
In this case, the nonlinear temperature is assumed to be distributed across the thickness of the plate in accordance with a one-dimensional steady state heat conduction equation. Thus, at a certain thickness, the following mathematical function is implemented to represent nonlinear temperature distribution [14, 60–62]:

$$-\frac{d}{dz}\left(k(z)\frac{dT}{dz} \right) = 0$$

(9.5)

Where $k(z)$ is defined by Equation 9.19.

Nomenclature for temperatures is presented in Figure 9.2. Now, implementing the thermal boundary conditions and continuity condition at the interfaces shown in Table 9.1, the differential Equation 9.5 is solved, and we get for (Figure 9.3):

- **S-FGM plate**

$$T_b = T_1 + (T_2 - T_1)\frac{1}{c_1}\left[\hbar_{zb}\left(1 + \sum_{i=1}^{5} \aleph^i{}_1 \left(\hbar_{zb} \right)^{ip} \right) \right] \quad \left(\hbar_0 \leq z < \hbar_m \right)$$

(9.6)

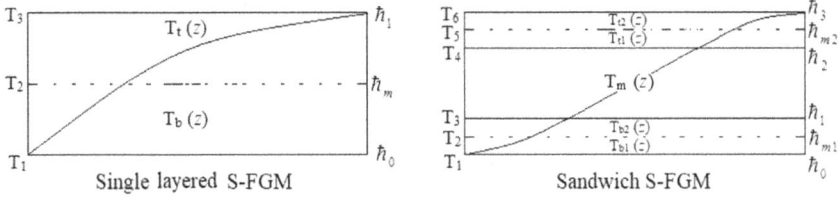

FIGURE 9.2   Thickness and temperature nomenclature [63].

TABLE 9.1
Boundary and Continuity Conditions

	Equations	
Condition	**S-FGM Plate**	**Sandwich S-FGM Plate**
Temperature at the boundaries and interfaces	$T\left(\hbar/2\right)=T_3$ and $T\left(-\hbar/2\right)=T_1$ $T_b\left(h_m\right)=T_t\left(h_m\right)=T_2$	$T\left(\hbar/2\right)=T_6$ and $T\left(-\hbar/2\right)=T_1$ $T_{b_1}\left(h_{m1}\right)=T_{b_2}\left(h_{m1}\right)=T_2,\ T_{b_2}\left(\hbar_1\right)=T_m\left(\hbar_1\right)=T_3,$ $T_m\left(\hbar_2\right)=T_{t_1}\left(\hbar_2\right)=T_4,\ T_{t_1}\left(h_{m2}\right)=T_{t_2}\left(h_{m2}\right)=T_5$
Continuity conditions	$\left.\dfrac{dT_b}{dz}\right|_{z=h_m}=\left.\dfrac{dT_t}{dz}\right|_{z=h_m}$	$\left.\dfrac{dT_{b_1}}{dz}\right|_{z=h_{m1}}=\left.\dfrac{dT_{b_2}}{dz}\right|_{z=h_{m1}},\ \left.\dfrac{dT_{b_2}}{dz}\right|_{z=h_1}=\left.\dfrac{dT_m}{dz}\right|_{z=h_1},$ $\left.\dfrac{dT_m}{dz}\right|_{z=h_2}=\left.\dfrac{dT_{t_1}}{dz}\right|_{z=h_2},\ \left.\dfrac{dT_{t_1}}{dz}\right|_{z=h_{m2}}=\left.\dfrac{dT_{t_2}}{dz}\right|_{z=h_{m2}}$

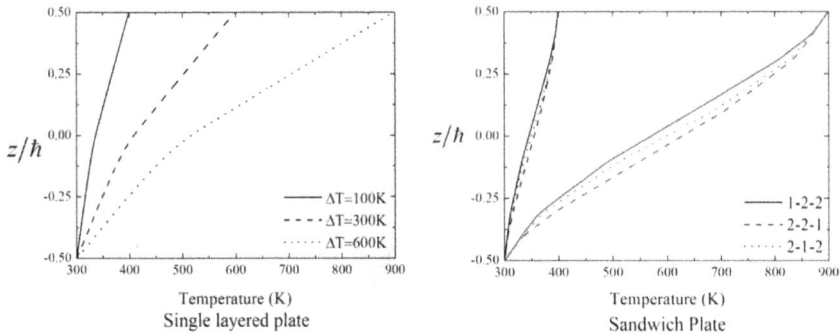

FIGURE 9.3 Temperature variation along the thickness (*black: $\Delta T=100$ K, red: $\Delta T=600$ K*) [63].

$$T_t = T_3 + (T_2 - T_3)\frac{1}{c_2}\left[\hbar_{zt}\left(1 + \sum_{i=1}^{5}\left(\aleph_4^i + \frac{\aleph_1^i}{(-1)^i}(\hbar_{zt})^{ip}\right)\right)\right] \quad (\hbar_m \leq z < \hbar_1) \quad (9.7)$$

where, $T_2 = \dfrac{T_1 c_2 (c_3 + 1) + T_3 c_1 (c_4 + c_5 + 1)}{c_2 (c_3 + 1) + c_1 (c_4 + c_5 + 1)}$,

$$\hbar_m = \frac{\hbar_0 + \hbar_1}{2}, \ \hbar_{zb} = \frac{z - \hbar_0}{\hbar_m - \hbar_0}, \hbar_{zt} = \frac{z - \hbar_1}{\hbar_m - \hbar_1}$$

• **Sandwich S-FGM plate**

$$T_{b_1} = T_1 + \frac{(T_2 - T_1)}{c_1}\left[\hbar_{z0}\left(1 + \sum_{i=1}^{5}\aleph_1^i(\hbar_{z0})^{ip}\right)\right] \quad (\hbar_0 \leq z < \hbar_{m1}) \quad\quad (9.8)$$

$$T_{b_2} = T_3 + \frac{(T_2 - T_3)}{c_2}\left[\hbar_{z1}\left(1 + \sum_{i=1}^{5}\left(\aleph_4^i + (\hbar_{z1})^{ip}\frac{\aleph_1^i}{(-1)^i}\right)\right)\right] \quad (\hbar_{m1} \leq z < \hbar_1) \quad (9.9)$$

$$T_m = T_3 + (T_3 - T_4)\frac{z - \hbar_1}{\hbar_1 - \hbar_2} \quad (\hbar_1 \leq z < \hbar_2) \quad\quad (9.10)$$

$$T_{t_1} = T_4 + \frac{(T_5 - T_4)}{c_2}\left[\hbar_{z2}\left(1 + \sum_{i=1}^{5}\left(\aleph_4^i + (\hbar_{z2})^{ip}\frac{\aleph_1^i}{(-1)^i}\right)\right)\right] \quad (\hbar_2 \leq z < \hbar_{m2}) \quad (9.11)$$

$$T_{t_2} = T_6 + \frac{(T_5 - T_6)}{c_1}\left[\hbar_{z3}\left(1 + \sum_{i=1}^{5}\aleph_1^i(\hbar_{z3})^{ip}\right)\right] \quad (\hbar_{m2} \leq z < \hbar_3) \quad\quad (9.12)$$

where, $\hbar_{m1} = \dfrac{\hbar_0 + \hbar_1}{2}, \ \hbar_{m2} = \dfrac{\hbar_2 + \hbar_3}{2}, \ \hbar_{z0} = \dfrac{z - \hbar_0}{\hbar_{m1} - \hbar_0}$,

$$\hbar_{z1} = \frac{z - \hbar_1}{\hbar_{m1} - \hbar_1}, \ \hbar_{z2} = \frac{z - \hbar_2}{\hbar_{m2} - \hbar_2}, \ \hbar_{z3} = \frac{z - \hbar_3}{\hbar_{m2} - \hbar_3}$$

$$T_2 = \frac{T_1\left(c_1 H_{f_2}\left(c_4+c_5+1\right)+c_2\left(c_3+1\right)\left(H_{f_1}+H_{f_2}\right)+2\left(c_3+1\right)\left(c_4+1\right)H_c\right)+T_6 c_1 H_{f_1}\left(c_4+c_5+1\right)}{\left(H_{f_1}+H_{f_2}\right)\left(c_1\left(c_4+c_5+1\right)+c_2\left(c_3+1\right)\right)+2\left(c_3+1\right)\left(c_4+1\right)H_c}$$

$$T_3 = \frac{T_1\left(H_{f_2}\left(c_2\left(c_3+1\right)+c_1\left(c_4+c_5+1\right)\right)+2\left(c_3+1\right)\left(c_4+1\right)H_c\right)+T_6 H_{f_1}\left(c_2\left(c_3+1\right)+c_1\left(c_4+c_5+1\right)\right)}{\left(H_{f_1}+H_{f_2}\right)\left(c_1\left(c_4+c_5+1\right)+c_2\left(c_3+1\right)\right)+2\left(c_3+1\right)\left(c_4+1\right)H_c}$$

$$T_4 = \frac{T_1 H_{f_2}\left(c_2\left(c_3+1\right)+c_1\left(c_4+c_5+1\right)\right)+T_6\left(2\left(c_3+1\right)\left(c_4+1\right)H_c+\left(c_1\left(c_4+c_5+1\right)+c_2\left(c_3+1\right)\right)H_{f_1}\right)}{\left(H_{f_1}+H_{f_2}\right)\left(c_1\left(c_4+c_5+1\right)+c_2\left(c_3+1\right)\right)+2\left(c_3+1\right)\left(c_4+1\right)H_c}$$

$$T_5 = \frac{T_1 c_1 H_{f_2}\left(c_4+c_5+1\right)+T_6\left(H_{f_1}c_1\left(c_4+c_5+1\right)+2\left(c_3+1\right)\left(c_4+1\right)H_c+c_2\left(c_3+1\right)\left(H_{f_1}+H_{f_2}\right)\right)}{\left(H_{f_1}+H_{f_2}\right)\left(c_1\left(c_4+c_5+1\right)+c_2\left(c_3+1\right)\right)+2\left(c_3+1\right)\left(c_4+1\right)H_c}$$

$$H_{f_1}=h_1-h_0,\ H_{f_2}=h_3-h_2,\ H_c=h_2-h_1$$

c_k's and $\aleph_i's$ are the constants and given as,

$$c_k = 1+\sum_{i=1}^{5}\aleph^i{}_k \quad \text{where,}\ k=1,2,3,4,5$$

$$\aleph^i{}_1 = (-1)^i\left(\frac{1}{2}\right)^i\left(\frac{k_{cm}}{k_m}\right)^i\left(\frac{1}{1+ip}\right),$$

$$\aleph^i{}_2 = \left(\frac{k_{cm}}{k_m}\right)^i(-1)^i\left(1+(-1)^i\left(\frac{1}{2}\right)^i\left(\frac{1}{1+ip}\right)\right)$$

$$\aleph^i{}_3 = (-1)^i\left(\frac{1}{2}\right)^i\left(\frac{k_{cm}}{k_m}\right)^i,\ \aleph^i{}_4 = (-1)^i\left(\frac{k_{cm}}{k_m}\right)^i,\ \aleph^i{}_5 = \left(\frac{1}{2}\right)^i\left(\frac{k_{cm}}{k_m}\right)^i$$

9.3 MATERIAL PROPERTIES AND CONSTITUTIVE RELATION

The structures are subjected to dynamic load and require more strength and stiffness that can be accomplished by implementing a sigmoid function-based FGM [22, 52–54, 64] when working under the same conditions as other micromechanics models (E- and P-FGM). In addition, P- and E-FGM are encountered with enormous stress concentrations across the interfaces where an abrupt change in material takes place [65]. To avoid this, a two-power law-based function is proposed, also known as S-FGM, by Chung and Chi [66] and expressed as:

$$V_t = \frac{1}{2}\left(\frac{z-\hbar_0}{\hbar_m-\hbar_0}\right)^p \quad \text{for}\ \hbar_0 \le z \le \hbar_m$$

$$V_t = 1-\frac{1}{2}\left(\frac{z-\hbar_1}{\hbar_i-\hbar_1}\right)^p \quad \text{for}\ \hbar_m \le z \le \hbar_1$$

(9.13)

By altering the sigmoid function, the volume fraction for a sandwich S-FGM plate is formulated in such a way that the continuousness of the material properties at the boundary is still preserved and is given as:

(a) Bottom layer:

$$V_t = \frac{1}{2}\left(\frac{z-\hbar_0}{\hbar_{m_1}-\hbar_0}\right)^p \quad \text{for } \hbar_0 \leq z \leq \hbar_{m_1}$$

$$V_t = 1 - \frac{1}{2}\left(\frac{z-\hbar_1}{\hbar_{m_1}-\hbar_1}\right)^p \quad \text{for } \hbar_{m_1} \leq z \leq \hbar_1 \tag{9.14}$$

(b) Core:

$$V_t = 1 \quad \text{for } \hbar_1 \leq z \leq \hbar_2 \tag{9.15}$$

(c) Top layer:

$$V_t = 1 - \frac{1}{2}\left(\frac{z-\hbar_2}{\hbar_{m_2}-\hbar_2}\right)^p \quad \text{for } \hbar_2 \leq z \leq \hbar_{m_2}$$

$$V_t = \frac{1}{2}\left(\frac{z-\hbar_3}{\hbar_{m_2}-\hbar_3}\right)^p \quad \text{for } \hbar_{m_2} \leq z \leq \hbar_3 \tag{9.16}$$

The parameters $\hbar_0, \hbar_1, \hbar_2, \hbar_3, \hbar_m, \hbar_{m_1}, \hbar_{m_2}$ are defined in Figure 9.2. The various configurations (symmetric and non-symmetric) for the sandwich plate are tabulated in Table 9.2.

TABLE 9.2
Plate Configurations

Configuration	Ω_1^*	Ω_2^*	Description
1-1-1	6	6	All the layers are of equal thickness.
2-1-2	10	10	The core is half the thickness of the top and bottom layers.
2-2-1	10/3	10	The core is equal in thickness to the top layer and double the thickness of the bottom layer.
2-1-1	4	∞	The core is equal in thickness to the bottom layer and half the thickness of the top layer.

* $\hbar_1 = -\hbar/\Omega_1, \hbar_2 = \hbar/\Omega_2$.

The estimation of material properties at a certain thickness across the sandwich plate for the ith layer with porosity effect is expressed as:

$$\wp^{(i)}(z) = \wp_c V_t^{(i)} + \wp_m \left(1 - V_t^{(i)}\right) - (\wp_c + \wp_m)P_0 \tag{9.17}$$

The effective material properties of the constituent material (ceramic/metal) are a function of temperature by the following functional relationship [67]:

$$\wp_{(c/m)} = \mathcal{M}_0 \left(\mathcal{M}_{-1} T_{(i)}^{-1} + 1 + \mathcal{M}_1 T_{(i)} + \mathcal{M}_2 T_{(i)}^{2} + \mathcal{M}_3 T_{(i)}^{3}\right) \tag{9.18}$$

In addition, it is deduced from calculation and the available literature that considering the porosity effect on thermal conductivity in Equation 9.5 is computationally expensive; therefore, the effective material property independent of porosity effect for thermal conductivity is expressed as:

$$k^{(i)}(z) = k_c V_t^{(i)} + k_m \left(1 - V_t^{(i)}\right) \tag{9.19}$$

Where superscript i represents the corresponding layer of the sandwich plate. $T_{(i)}$ and $\mathcal{M}$ are the temperature function across the thickness and the material constant (Table 9.3), respectively. Also, material properties at room temperature (300 K) are evaluated from Equation 9.18.

The reduction in volume fraction of the constituent materials is represented by P_0 and equal to either $P - 1$ or $P - 2$ or $P - 3$ depending upon the type of porosity distribution assumed from Equations 9.1-9.3. Also, the degradation of material properties due to porosity and thermal effect using Equations 9.5.a and 9.5.b is presented in Figure 9.4.

The stress-strain thermo-mechanical constitutive relation can be expressed for an FGM plate as:

$$\begin{bmatrix} \sigma_{xx} \\ \sigma_{yy} \\ \tau_{yz} \\ \tau_{xz} \\ \tau_{xy} \end{bmatrix} = \begin{bmatrix} C_{11} & C_{12} & 0 & 0 & 0 \\ C_{12} & C_{22} & 0 & 0 & 0 \\ 0 & 0 & C_{44} & 0 & 0 \\ 0 & 0 & 0 & C_{55} & 0 \\ 0 & 0 & 0 & 0 & C_{66} \end{bmatrix} \begin{Bmatrix} \varepsilon_{xx} - \alpha \Delta T \\ \varepsilon_{yy} - \alpha \Delta T \\ \gamma_{yz} \\ \gamma_{xz} \\ \gamma_{xy} \end{Bmatrix} \tag{9.20}$$

$$C_{11} = C_{22} = \frac{E(z,T)}{1 - v^2}, \quad C_{12} = \frac{vE(z,T)}{1 - v^2}, \quad C_{44} = C_{55} = C_{66} = \frac{E(z,T)}{2(1 + v)}$$

Where $\Delta T = T(z) - T_0$, and $T(z)$ represents the temperature distribution across the thickness. $E(z,T)$ and $\alpha(z,T)$ are the Young's modulus and thermal expansion coefficient, respectively. v is a Poisson's ratio which remains constant. The reference temperature (T_0) is defined as temperature with no thermal strains.

TABLE 9.3
Material Constant as Function of Temperature [68]

Materials	Properties	Temperature Dependent (TD)					Temperature Independent (TID)
		M_{-1}	M_0	M_1	M_2	M_3	$M^{(300K)}$
ZrO_2	E (Pa)	0	244.27×10^9	-1.371×10^{-3}	1.214×10^{-6}	-3.681×10^{-10}	168.063×10^9
	α (1/K)	0	12.766×10^{-6}	-1.491×10^{-3}	1.006×10^{-5}	-6.778×10^{-11}	18.591×10^{-6}
	k (W/mK)	0	0	0	0	0	1.77
	ρ (kg/m³)	0	0	0	0	0	3000
Si_3N_4	E (Pa)	0	348.43×10^9	-3.070×10^{-4}	2.160×10^{-7}	-8.946×10^{-11}	322.2715×10^9
	α (1/K)	0	5.8723×10^{-6}	9.095×10^{-4}	0	0	7.4746×10^{-6}
	k (W/mK)	0	0	0	0	0	8.828
	ρ (kg/m³)	0	0	0	0	0	2370
SUS304	E (Pa)	0	201.04×10^9	3.079×10^{-4}	-6.534×10^{-7}	0	207.7877×10^9
	α (1/K)	0	12.330×10^{-6}	8.086×10^{-4}	0	0	15.321×10^{-6}
	k (W/mK)	0	0	0	0	0	12.14
	ρ (kg/m³)	0	0	0	0	0	8166

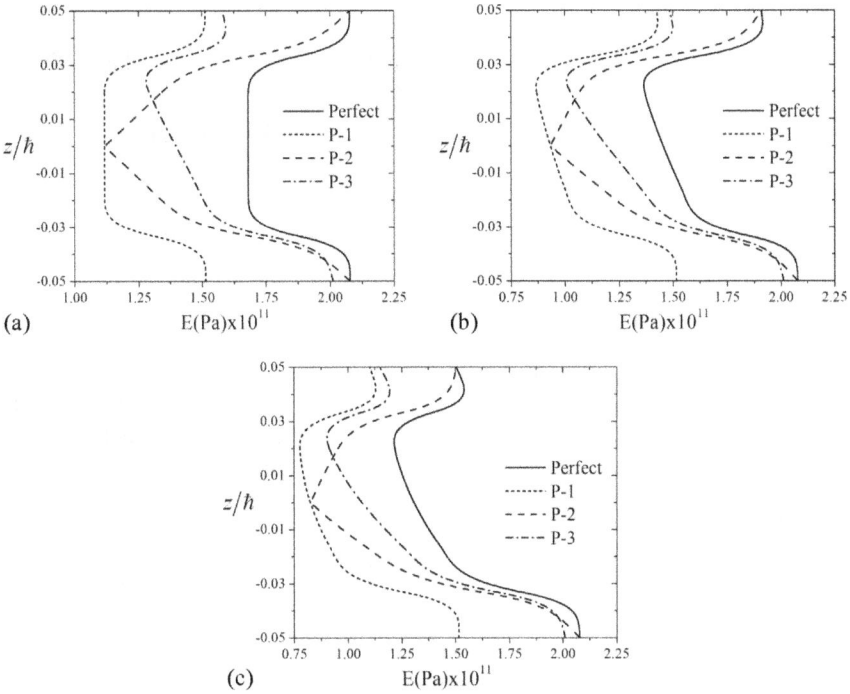

(a)

(b)

(c)

FIGURE 9.4 Variation of the modulus of elasticity in the thickness direction of the (1-1-1) porous sandwich plate at a temperature difference of (a) 0 K, (b) 300 K, (c) 600K.

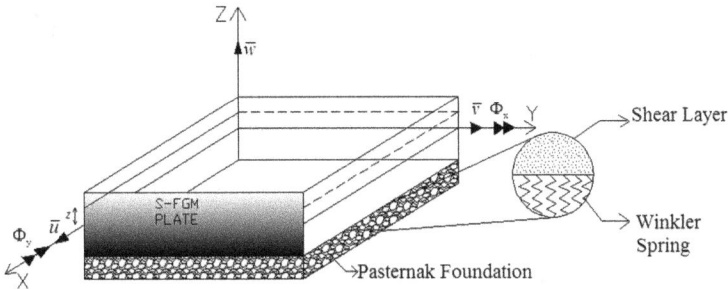

FIGURE 9.5 Nomenclature for the displacement field.

9.4 THEORETICAL FORMULATION

The plate considered in the present study has dimensions of $a \times b \times \hbar$ in a rectangular co-ordinate system. In addition, the stiffness of the plate has been increased by assuming the plate to be placed on a Pasternak foundation which will offer stiffness both in the normal direction (Winkler spring) and shear effect (Pasternak) as depicted in Figure 9.5.

9.4.1 KINEMATICS

The displacement field (u, v, w) for the present formulation is expressed as [69]:

$$u(x,y,z,t) = \bar{u}(x,y,t) - z\left(\frac{\partial \bar{w}}{\partial x}\right) + \left(\lambda(z) + z\vartheta\right)\Phi_x(x,y,t)$$

$$v(x,y,z,t) = \bar{v}(x,y,t) - z\left(\frac{\partial \bar{w}}{\partial y}\right) + \left(\lambda(z) + z\vartheta\right)\Phi_y(x,y,t) \qquad (9.21)$$

$$w(x,y,z,t) = \bar{w}(x,y,t)$$

Where $\lambda(z) = \tanh^{-1}\left(rz/\hbar\right)$ and $\vartheta = -(r/\hbar)/(1-r^2/4)$, $r=0.088$. The displacement of the mid-surface in the plane of the plate is $\bar{u}, \bar{v}$ and out-of-the plane is $\bar{w}$. The rotations of the mid-surface or the coefficients of the shear function are Φ_x and Φ_y.

The constitutive strain-displacement relation instigating the von Karman nonlinearity in a concise form can be expressed as:

$$\varepsilon = \varepsilon^{(0)} + z\varepsilon_b^{(1)} + f(z)\varepsilon_s^{(1)} \qquad (9.22)$$

$$\gamma = f'(z)\gamma_s^{(0)} \qquad (9.23)$$

In expanded form, Equation 9.22 and Equation 9.23 can be expressed as:

$$\varepsilon^{(0)} = \left\{\varepsilon^{(0)}{}_{xx}, \varepsilon^{(0)}{}_{yy}, \gamma^{(0)}{}_{xy}\right\}$$
$$= \left\{\frac{\partial \bar{u}}{\partial x} + \frac{1}{2}\left(\frac{\partial \bar{w}}{\partial x}\right)^2, \frac{\partial \bar{v}}{\partial y} + \frac{1}{2}\left(\frac{\partial \bar{w}}{\partial y}\right)^2, \frac{\partial \bar{u}}{\partial y} + \frac{\partial \bar{v}}{\partial x} + \left(\frac{\partial \bar{w}}{\partial x}\right)\left(\frac{\partial \bar{w}}{\partial y}\right)\right\} \qquad (9.24)$$

$$\varepsilon_b^{(1)} = \left\{\varepsilon_b^{(1)}{}_{xx}, \varepsilon_b^{(1)}{}_{yy}, \gamma_b^{(1)}{}_{xy}\right\} = \left\{-\frac{\partial^2 \bar{w}}{\partial x^2}, -\frac{\partial^2 \bar{w}}{\partial y^2}, -2\frac{\partial^2 \bar{w}}{\partial x \partial y}\right\} \qquad (9.25)$$

$$\varepsilon_s^{(1)} = \left\{\varepsilon_s^{(1)}{}_{xx}, \varepsilon_s^{(1)}{}_{yy}, \gamma_s^{(1)}{}_{xy}\right\} = \left\{\frac{\partial \Phi_x}{\partial x}, \frac{\partial \Phi_y}{\partial y}, \frac{\partial \Phi_x}{\partial y} + \frac{\partial \Phi_y}{\partial x}\right\} \qquad (9.26)$$

$$\gamma_s^{(0)} = \left\{\gamma^{(0)}{}_{xz}, \gamma^{(0)}{}_{yz}\right\} = \left\{\Phi_x, \Phi_y\right\} \qquad (9.27)$$

9.4.2 ENERGY EQUATIONS

The variational form of the strain energy is expressed as,

$$\delta U = \iiint \left[\sigma_{xx}\delta\varepsilon_{xx} + \sigma_{yy}\delta\varepsilon_{yy} + 2(\sigma_{xy}\delta\varepsilon_{xy} + \sigma_{yz}\delta\varepsilon_{yz} + \sigma_{zx}\delta\varepsilon_{zx})\right] dV \qquad (9.28)$$

In the form of stress resultants, the above equation is re-written by substituting Equation 9.20 and Equations 9.24–9.27 into Equation 9.28, and expressed as:

$$
\delta U = \int_A \begin{bmatrix} \left(N_{xx}\delta\varepsilon^{(0)}_{xx} + N_{yy}\delta\varepsilon^{(0)}_{yy} + N_{xy}\delta\gamma^{(0)}_{xy} \right) \\[4pt] + \left(M^b_{xx}\delta\varepsilon_b^{(1)}_{xx} + M^b_{yy}\delta\varepsilon_b^{(1)}_{yy} + M^b_{xy}\delta\gamma_b^{(1)}_{xy} \right) \\[4pt] + \left(M^s_{xx}\delta\varepsilon_s^{(1)}_{xx} + M^s_{yy}\delta\varepsilon_s^{(1)}_{yy} + M^s_{xy}\delta\gamma_s^{(1)}_{xy} \right) \\[4pt] + \left(M^q_{xx}\delta\Phi_x + M^q_{yy}\delta\Phi_y \right) + \vartheta\left(Q_x\delta\Phi_x + Q_y\delta\Phi_y \right) \end{bmatrix} dxdy \qquad (9.29)
$$

where

$$
\begin{Bmatrix} N_{xx} & M^b_{xx} & M^s_{xx} \\ N_{yy} & M^b_{yy} & M^s_{yy} \\ N_{xy} & M^b_{xy} & M^s_{xy} \end{Bmatrix} = \int_{-\hbar/2}^{\hbar/2} (1,z,\lambda(z)) \begin{Bmatrix} \sigma_{xx} \\ \sigma_{yy} \\ \tau_{xy} \end{Bmatrix} dz \qquad (9.30)
$$

$$
\begin{Bmatrix} Q_x & M^q_{xx} \\ Q_y & M^q_{yy} \end{Bmatrix} = \int_{-\hbar/2}^{\hbar/2} (1,\lambda'(z)) \begin{Bmatrix} \tau_{xz} \\ \tau_{yz} \end{Bmatrix} dz \qquad (9.31)
$$

Further, the stress resultants are expanded by substituting Equation 9.20 into Equations 9.30–9.31 and are expressed as:

$$
\begin{bmatrix} N_{xx} & M^b_{xx} & M^s_{xx} \\ N_{yy} & M^b_{yy} & M^s_{yy} \\ N_{xy} & M^b_{xy} & M^s_{xy} \end{bmatrix}
$$

$$
= \frac{1}{1-v^2} \begin{bmatrix} \xi_{11} & \xi_{12} & \xi_{13} \\ \xi_{21} & \xi_{22} & \xi_{23} \\ \left(\dfrac{1-v}{2}\right)\xi_{31} & \left(\dfrac{1-v}{2}\right)\xi_{32} & \left(\dfrac{1-v}{2}\right)\xi_{33} \end{bmatrix} \begin{bmatrix} A & B & F \\ B & D & H \\ F & H & J \end{bmatrix} \qquad (9.32)
$$

$$
- \frac{1}{1-v} \begin{bmatrix} A^T & B^T & F^T \\ A^T & B^T & F^T \\ 0 & 0 & 0 \end{bmatrix}
$$

$$
\begin{Bmatrix} Q_x \\ Q_y \end{Bmatrix} = \frac{(9A+K)}{2(1+v)} \begin{Bmatrix} \gamma^{(0)}_{xz} \\ \gamma^{(0)}_{yz} \end{Bmatrix} \qquad (9.33)
$$

$$\begin{Bmatrix} M^q{}_{xx} \\ M^q{}_{yy} \end{Bmatrix} = \frac{(9K+L)}{2(1+v)} \begin{Bmatrix} \gamma^{(0)}{}_{xz} \\ \gamma^{(0)}{}_{yz} \end{Bmatrix}$$

(9.34)

where

$$\xi_{11} = \left(\varepsilon^{(0)}{}_{xx} + v\varepsilon^{(0)}{}_{yy} \right), \; \xi_{12} = \varepsilon_b{}^{(1)}{}_{xx} + 9\varepsilon_s{}^{(1)}{}_{xx} + v\left(\varepsilon_b{}^{(1)}{}_{yy} + 9\varepsilon_s{}^{(1)}{}_{yy} \right),$$
$$\xi_{13} = \varepsilon_s{}^{(1)}{}_{xx} + v\varepsilon_s{}^{(1)}{}_{yy}$$

$$\xi_{21} = \left(\varepsilon^{(0)}{}_{yy} + v\varepsilon^{(0)}{}_{xx} \right), \; \xi_{22} = \varepsilon_b{}^{(1)}{}_{yy} + 9\varepsilon_s{}^{(1)}{}_{yy} + v\left(\varepsilon_b{}^{(1)}{}_{xx} + 9\varepsilon_s{}^{(1)}{}_{xx} \right),$$
$$\xi_{23} = \varepsilon_s{}^{(1)}{}_{yy} + v\varepsilon_s{}^{(1)}{}_{xx}$$

$$\xi_{31} = \gamma^{(0)}{}_{xy}, \; \xi_{32} = \gamma_b{}^{(1)}{}_{xy} + 9\gamma_s{}^{(1)}{}_{xy}, \; \xi_{33} = \gamma_s{}^{(1)}{}_{xy}$$

The plate stiffnesses are defined as,

$$\begin{Bmatrix} A, B, D, F, H, \\ J, K, L \end{Bmatrix} = \int_{-h/2}^{h/2} E(z) \begin{Bmatrix} 1, z, z^2, \lambda(z), z\lambda(z), \\ \lambda(z)^2, \lambda'(z), \lambda'(z)^2 \end{Bmatrix} dz$$

(9.35)

$$\left\{ A^T, B^T, F^T \right\} = \int_{h_{i-1}}^{h_i} \alpha(z,T) E(z,T) \Delta T \left\{ 1, z, \lambda(z) \right\} dz \quad \text{for } i = 1,2,3,4,5.$$

(9.36)

$$\left\{ N^T, M_b{}^T, M_s{}^T \right\} = \frac{1}{1-v} \int_{h_{i-1}}^{h_i} \alpha(z,T) E(z,T) \Delta T \left\{ 1, z, \lambda(z) \right\} dz$$

(9.37)

The variational form of work done due to load applied on the surface of the plate can be expressed as:

$$\delta W = -\iint q_z \delta \bar{w} \, dx dy$$

(9.38)

The variational form of the kinetic energy is expressed as,

$$\delta K = \int_A \int_{-h/2}^{h/2} \rho \left(\dot{u} \, \delta \dot{u} + \dot{v} \, \delta \dot{v} + \dot{w} \, \delta \dot{w} \right) dz \, dA$$

(9.39)

Substituting Equation 9.21 into Equation 9.39

$$\delta K = \int_A \begin{bmatrix} I_0\left(\dot{u}\,\delta\dot{u} + \dot{v}\,\delta\dot{v} + \dot{w}\,\delta\dot{w}\right) \\[2mm] -I_1\left(\dot{u}\,\dfrac{\partial\delta\dot{w}}{\partial x} + \dfrac{\partial\dot{w}}{\partial x}\,\delta\dot{u} + \dot{v}\,\dfrac{\partial\delta\dot{w}}{\partial y} + \dfrac{\partial\dot{w}}{\partial y}\,\delta\dot{v}\right) \\[2mm] +I_2\left(\dfrac{\partial\dot{w}}{\partial x}\dfrac{\partial\delta\dot{w}}{\partial x} + \dfrac{\partial\dot{w}}{\partial y}\dfrac{\partial\delta\dot{w}}{\partial y}\right) \\[2mm] +J_1\left(\dot{u}\,\delta\dot{\Phi}_x + \dot{\Phi}_x\,\delta\dot{u} + \dot{v}\,\delta\dot{\Phi}_y + \dot{\Phi}_y\,\delta\dot{v}\right) \\[2mm] -J_2\left(\dfrac{\partial\dot{w}}{\partial x}\,\delta\dot{\Phi}_x + \dot{\Phi}_x\,\dfrac{\partial\delta\dot{w}}{\partial x} + \dfrac{\partial\dot{w}}{\partial y}\,\delta\dot{\Phi}_y + \dot{\Phi}_y\,\dfrac{\partial\delta\dot{w}}{\partial y}\right) \\[2mm] +J_3\left(\dot{\Phi}_x\,\delta\dot{\Phi}_x + \dot{\Phi}_y\,\delta\dot{\Phi}_y\right) \end{bmatrix} dxdy \qquad (9.40)$$

where $\left\{I_0, I_1, I_2, J_1, J_2, J_3\right\} = \displaystyle\int_{-h/2}^{h/2} \rho\left\{1, z, z^2, f(z), zf(z), f(z)^2\right\} dz.$

The variational form of the strain energy of the Pasternak foundation is expressed as,

$$\delta V = \iint \hat{N}_e \delta\overline{w}\, dxdy \qquad (9.41)$$

where $\hat{N}_e = \left(K_w - K_P\nabla^2\right)\overline{w}$, and K_w and K_p are the Winkler and Pasternak stiffness coefficients.

9.4.3 GOVERNING EQUATIONS

In the present study, Hamilton's principle is employed to derive the governing differential equations and is expressed as:

$$\int_{t_1}^{t_2} \left(\delta U + \delta V + \delta W - \delta K\right) dt = 0 \qquad (9.42)$$

To arrive at governing differential equations, Equations 9.29, 9.38, 9.40, and 9.41 are substituted into Equation 9.42. Further, applying integration by parts and collecting the coefficients of in-plane displacements and rotations $\left(\overline{u}, \overline{v}, \overline{w}, \Phi_x, \Phi_y\right)$ term by term. This will result in the following equations:

$$\delta\overline{u}: \quad \frac{\partial N_{xx}}{\partial x} + \frac{\partial N_{xy}}{\partial y} = I_0\ddot{\overline{u}} - I_1\frac{\partial\ddot{\overline{w}}}{\partial x} + J_1\ddot{\Phi}_x \qquad (9.43)$$

$$\delta \bar{v} : \frac{\partial N_{yy}}{\partial y} + \frac{\partial N_{xy}}{\partial x} = I_0 \ddot{\bar{v}} - I_1 \frac{\partial \ddot{\bar{w}}}{\partial y} + J_1 \ddot{\Phi}_y \tag{9.44}$$

$$\delta \bar{w} : \left\{ \begin{array}{l} \left(\dfrac{\partial^2 M^b_{xx}}{\partial x^2} + \dfrac{\partial^2 M^b_{yy}}{\partial y^2} + 2 \dfrac{\partial^2 M^b_{xy}}{\partial x \partial y} \right) + q_z - N_e \\[2ex] + \dfrac{\partial}{\partial x} \left(N_{xx} \dfrac{\partial \bar{w}}{\partial x} + N_{xy} \dfrac{\partial \bar{w}}{\partial y} \right) \\[2ex] + \dfrac{\partial}{\partial y} \left(N_{yy} \dfrac{\partial \bar{w}}{\partial y} + N_{xy} \dfrac{\partial \bar{w}}{\partial x} \right) \end{array} \right\} = \left\{ \begin{array}{l} I_0 \ddot{\bar{w}} + I_1 \left(\dfrac{\partial \ddot{u}}{\partial x} + \dfrac{\partial \ddot{v}}{\partial y} \right) \\[2ex] - I_2 \left(\dfrac{\partial^2 \ddot{\bar{w}}}{\partial x^2} + \dfrac{\partial^2 \ddot{\bar{w}}}{\partial y^2} \right) \\[2ex] + J_2 \left(\dfrac{\partial \ddot{\Phi}_x}{\partial x} + \dfrac{\partial \ddot{\Phi}_y}{\partial y} \right) \end{array} \right\} \tag{9.45}$$

$$\delta \Phi_x : \frac{\partial M^s_{xx}}{\partial x} + \vartheta \frac{\partial M^b_{xx}}{\partial x} + \frac{\partial M^s_{xy}}{\partial y} + \vartheta \frac{\partial M^b_{xy}}{\partial y} - M^q_{xx} - \vartheta Q_x$$

$$= J_1 \ddot{u} - J_2 \frac{\partial \ddot{\bar{w}}}{\partial x} + J_3 \ddot{\Phi}_x \tag{9.46}$$

$$\delta \Phi_y : \frac{\partial M^s_{yy}}{\partial y} + \vartheta \frac{\partial M^b_{yy}}{\partial y} + \frac{\partial M^s_{yx}}{\partial x} + \vartheta \frac{\partial M^b_{yx}}{\partial x} - M^q_{yy} - \vartheta Q_y$$

$$= J_1 \ddot{v} - J_2 \frac{\partial \ddot{\bar{w}}}{\partial y} + J_3 \ddot{\Phi}_y \tag{9.47}$$

9.4.3.1 Airy's Function and Strain Compatibility Equation

Airy's function $\phi(x, y, t)$ is introduced to solve nonlinearity existent in the governing differential equations and given as [70–72]:

$$N_x = \frac{\partial^2 \phi}{\partial y^2}, N_y = \frac{\partial^2 \phi}{\partial x^2}, N_{xy} = -\frac{\partial^2 \phi}{\partial x \partial y} \tag{9.48}$$

The strain compatibility equation for the problem accounting for geometric nonlinearity can be expressed as:

$$\frac{\partial^2 \varepsilon^{(0)}_{xx}}{\partial y^2} + \frac{\partial^2 \varepsilon^{(0)}_{yy}}{\partial x^2} - \frac{\partial^2 \gamma^{(0)}_{xy}}{\partial x \partial y} = \left(\frac{\partial^2 \bar{w}}{\partial x \partial y} \right)^2 - \frac{\partial^2 \bar{w}}{\partial x^2} \frac{\partial^2 \bar{w}}{\partial y^2} \tag{9.49}$$

Consequently, linear strain from Equation 9.32, in terms of stress resultants and higher order strain, is expressed as:

$$\varepsilon^{(0)}_{xx} = \frac{1}{A} \left(N_{xx} - \nu N_{yy} - B \left(\varepsilon_b^{(1)}{}_{xx} + \vartheta \varepsilon_s^{(1)}{}_{xx} \right) - F \varepsilon_s^{(1)}{}_{xx} + A^T \right) \tag{9.50}$$

$$\varepsilon^{(0)}{}_{yy} = \frac{1}{A}\left(N_{yy} - vN_{xx} - B\left(\varepsilon_b^{(1)}{}_{yy} + 9\varepsilon_s^{(1)}{}_{yy}\right) - F\varepsilon_s^{(1)}{}_{yy} + A^T\right) \tag{9.51}$$

$$\gamma^{(0)}{}_{xy} = \frac{1}{A}\left(2(1+v)N_{xy} - B\left(\gamma_b^{(1)}{}_{xy} + 9\gamma_s^{(1)}{}_{xy}\right) - F\gamma_s^{(1)}{}_{xy}\right) \tag{9.52}$$

The higher order strains from Equations 9.24–9.27 are substituted into Equations 9.50–9.52 and the resulting expression is substituted into Equation 9.49. The strain compatibility is then expressed as:

$$\frac{\partial^4 \phi}{\partial y^4} + 2\frac{\partial^4 \phi}{\partial x^2 \partial y^2} + \frac{\partial^4 \phi}{\partial x^4} + \left(\frac{\partial^2 A^T}{\partial x^2} + \frac{\partial^2 A^T}{\partial y^2}\right) = A\left(\left(\frac{\partial^2 \overline{w}}{\partial x \partial y}\right)^2 - \frac{\partial^2 \overline{w}}{\partial x^2}\frac{\partial^2 \overline{w}}{\partial y^2}\right) \tag{9.53}$$

9.4.3.2 Equilibrium Equations

Equation 9.48 is now substituted into Equations 9.43 and 9.44 and the resulting expression is substituted back into Equations 9.45, 9.46, and 9.47. We get

$$\left\{\begin{matrix} \left(\dfrac{\partial^2 M^b{}_{xx}}{\partial x^2} + \dfrac{\partial^2 M^b{}_{yy}}{\partial y^2} + 2\dfrac{\partial^2 M^b{}_{xy}}{\partial x \partial y} + q_z\right) \\[2mm] + \dfrac{\partial}{\partial x}\left(N_{xx}\dfrac{\partial \overline{w}}{\partial x} + N_{xy}\dfrac{\partial \overline{w}}{\partial y}\right) \\[2mm] + \dfrac{\partial}{\partial y}\left(N_{yy}\dfrac{\partial \overline{w}}{\partial y} + N_{xy}\dfrac{\partial \overline{w}}{\partial x}\right) + q_z - \hat{N}_e \end{matrix}\right\} = \left\{\begin{matrix} I_0\ddot{w} + \left(\dfrac{I_1^2}{I_0} - I_2\right)\left(\dfrac{\partial^2 \ddot{w}}{\partial x^2} + \dfrac{\partial^2 \ddot{w}}{\partial y^2}\right) \\[2mm] + \left(J_2 - \dfrac{I_1 J_1}{I_0}\right)\left(\dfrac{\partial \ddot{\Phi}_x}{\partial x} + \dfrac{\partial \ddot{\Phi}_y}{\partial y}\right) \end{matrix}\right\} \tag{9.54}$$

$$\left\{\begin{matrix} \dfrac{\partial M^s{}_{xx}}{\partial x} + \Delta\dfrac{\partial M^b{}_{xx}}{\partial x} + \dfrac{\partial M^s{}_{xy}}{\partial y} \\[2mm] + \Delta\dfrac{\partial M^b{}_{xy}}{\partial y} - M^q{}_{xx} - \Delta Q_x \end{matrix}\right\} = \left(J_3 - \dfrac{J_1^2}{I_0}\right)\ddot{\Phi}_x + \left(\dfrac{J_1 I_1}{I_0} - J_2\right)\dfrac{\partial \ddot{w}}{\partial x} \tag{9.55}$$

$$\left\{\begin{matrix} \dfrac{\partial M^s{}_{yy}}{\partial y} + \Delta\dfrac{\partial M^b{}_{yy}}{\partial y} + \dfrac{\partial M^s{}_{yx}}{\partial x} \\[2mm] + \Delta\dfrac{\partial M^b{}_{yx}}{\partial x} - M^q{}_{yy} - \Delta Q_y \end{matrix}\right\} = \left(J_3 - \dfrac{J_1^2}{I_0}\right)\ddot{\Phi}_y + \left(\dfrac{J_1 I_1}{I_0} - J_2\right)\dfrac{\partial \ddot{w}}{\partial y} \tag{9.56}$$

The governing equation in terms of displacements is obtained by substituting Equations 9.24–9.27 into the stress resultant equation (Equations 9.32–9.34) and the resulting stress resultants are substituted into Equations 9.54–9.56; we get:

$$\left\{ \begin{array}{l} \ell_{11}\left(\bar{w}\right)+\ell_{12}\left(\Phi_x\right)+\ell_{13}\left(\Phi_y\right) \\ +q_z-\hat{N}_e+\tilde{\ell} \end{array} \right\}$$

$$=\left\{ \begin{array}{l} I_0\ddot{\bar{w}}+\left(\dfrac{I_1^2}{I_0}-I_2\right)\left(\dfrac{\partial^2\ddot{\bar{w}}}{\partial x^2}+\dfrac{\partial^2\ddot{\bar{w}}}{\partial y^2}\right)+\left(J_2-\dfrac{I_1 J_1}{I_0}\right)\left(\dfrac{\partial\ddot{\Phi}_x}{\partial x}+\dfrac{\partial\ddot{\Phi}_y}{\partial y}\right) \\ +\dfrac{1}{1-\nu}\left(\left(\dfrac{\partial^2 B^T}{\partial x^2}+\dfrac{\partial^2 B^T}{\partial y^2}\right)-\dfrac{BF}{A}\left(\dfrac{\partial^2 A^T}{\partial x^2}+\dfrac{\partial^2 A^T}{\partial y^2}\right)\right) \end{array} \right\} \tag{9.57}$$

$$\ell_{21}\left(\bar{w}\right)+\ell_{22}\left(\Phi_x\right)+\ell_{23}\left(\Phi_y\right)$$

$$=\left\{ \begin{array}{l} \left(J_3-\dfrac{J_1^2}{I_0}\right)\ddot{\Phi}_x+\left(\dfrac{J_1 I_1}{I_0}-J_2\right)\dfrac{\partial\ddot{\bar{w}}}{\partial x} \\ +\dfrac{1}{1-\nu}\left(\left(\dfrac{\partial F^T}{\partial x}+\dfrac{\partial B^T}{\partial x}\vartheta\right)-\dfrac{F+B\vartheta}{A}\dfrac{\partial A^T}{\partial x}\right) \end{array} \right\} \tag{9.58}$$

$$\ell_{31}\left(\bar{w}\right)+\ell_{32}\left(\Phi_x\right)+\ell_{33}\left(\Phi_y\right)$$

$$=\left\{ \begin{array}{l} \left(J_3-\dfrac{J_1^2}{I_0}\right)\ddot{\Phi}_y+\left(\dfrac{J_1 I_1}{I_0}-J_2\right)\dfrac{\partial\ddot{\bar{w}}}{\partial y} \\ +\dfrac{1}{1-\nu}\left(\left(\dfrac{\partial F^T}{\partial y}+\dfrac{\partial B^T}{\partial y}\vartheta\right)-\dfrac{F+B\vartheta}{A}\dfrac{\partial A^T}{\partial y}\right) \end{array} \right\} \tag{9.59}$$

where ℓ_{ij} and $\tilde{\ell}$ are given as

$$\ell_{11}=\dfrac{1}{1-\nu^2}\left(\dfrac{B^2}{A}-D\right)\left(\dfrac{\partial^4\bar{w}}{\partial x^4}+\dfrac{\partial^4\bar{w}}{\partial y^4}+2\dfrac{\partial^4\bar{w}}{\partial x^2\partial y^2}\right)$$

$$\ell_{12}=-\dfrac{1}{1-\nu^2}\left(\dfrac{BF}{A}+\dfrac{B^2\vartheta}{A}-\left(D\vartheta+H\right)\right)\left(\dfrac{\partial^3\Phi_x}{\partial x^3}+\dfrac{\partial^3\Phi_x}{\partial x\partial y^2}\right)$$

$$\ell_{13}=-\dfrac{1}{1-\nu^2}\left(\dfrac{BF}{A}+\dfrac{B^2\vartheta}{A}-\left(D\vartheta+H\right)\right)\left(\dfrac{\partial^3\Phi_y}{\partial y^3}+\dfrac{\partial^3\Phi_y}{\partial y\partial x^2}\right)$$

$$\tilde{\ell}=\dfrac{\partial^2\phi}{\partial y^2}\dfrac{\partial^2\bar{w}}{\partial x^2}-2\dfrac{\partial^2\phi}{\partial x\partial y}\dfrac{\partial^2\bar{w}}{\partial x\partial y}+\dfrac{\partial^2\bar{w}}{\partial y^2}\dfrac{\partial^2\phi}{\partial x^2}$$

$$\ell_{21}=\dfrac{1}{\left(1-\nu^2\right)}\left(\dfrac{BF}{A}+\dfrac{B^2\vartheta}{A}-\left(D\vartheta+H\right)\right)\left(\dfrac{\partial^3\bar{w}}{\partial x^3}+\dfrac{\partial^3\bar{w}}{\partial x\partial y^2}\right)$$

$$\ell_{22}=\left(J-\dfrac{F^2}{A}+\vartheta\left(H-\dfrac{BF}{A}\right)-\vartheta\left(\dfrac{BF}{A}+\dfrac{B^2\vartheta}{A}-\left(D\vartheta+H\right)\right)\right)$$

$$\left(\dfrac{1}{\left(1-\nu^2\right)}\dfrac{\partial^2\Phi_x}{\partial x^2}+\dfrac{1}{2(1+\nu)}\dfrac{\partial^2\Phi_x}{\partial y^2}\right)-\dfrac{1}{2(1+\nu)}\left(A\vartheta^2+2\vartheta K+L\right)\Phi_x$$

$$\ell_{23} = \frac{1}{2(1-\nu)} \left[\begin{array}{c} \left(J - \dfrac{F^2}{A} + \vartheta \left(H - \dfrac{BF}{A} \right) \right. \\ \left. - \vartheta \left(\dfrac{BF}{A} + \dfrac{B^2 \vartheta}{A} - (D\vartheta + H) \right) \right) \end{array} \right] \frac{\partial^2 \Phi_y}{\partial x \partial y}$$

$$\ell_{31} = \frac{1}{(1-\nu^2)} \left(\frac{B^2 \vartheta}{A} + \frac{BF}{A} - D\vartheta - H \right) \left(\frac{\partial^3 \overline{w}}{\partial y^3} + \frac{\partial^3 \overline{w}}{\partial x^2 \partial y} \right)$$

$$\ell_{32} = \frac{1}{2(1-\nu)} \left[\begin{array}{c} \left(J - \dfrac{F^2}{A} + \vartheta \left(H - \dfrac{BF}{A} \right) \right. \\ \left. - \vartheta \left(\dfrac{BF}{A} + \dfrac{B^2 \vartheta}{A} - (D\vartheta + H) \right) \right) \end{array} \right] \frac{\partial^2 \Phi_x}{\partial x \partial y}$$

$$\ell_{33} = \left(J - \frac{F^2}{A} + \vartheta \left(H - \frac{BF}{A} \right) - \vartheta \left(\frac{BF}{A} + \frac{B^2 \vartheta}{A} - (D\vartheta + H) \right) \right)$$

$$\left(\frac{1}{(1-\nu^2)} \frac{\partial^2 \Phi_y}{\partial y^2} + \frac{1}{2(\nu+1)} \frac{\partial^2 \Phi_y}{\partial x^2} \right) - \frac{1}{2(\nu+1)} \left(A\vartheta^2 + 2K\vartheta + L \right) \Phi_y$$

9.5　SOLUTION PROCEDURE

Consider a simply supported sandwich rectangular plate with all edges immovable. The plate is assumed to be resting on a Pasternak foundation. The boundary conditions for all edges simply supported can be expressed as:

Boundary conditions parallel to x-axis:

$$\overline{w} = \overline{v} = \Phi_x = M_y = 0, N_y = N_{y0}$$

Boundary conditions parallel to y-axis:

$$\overline{w} = \overline{u} = \Phi_y = M_x = 0, N_x = N_{x0}$$

Where N_{x0} and N_{y0} represent equivalent axial compressive loads along the x- and y-axes, respectively, and prevent the respective edges from moving.

9.5.1　ASSUMED SOLUTIONS AND TRANSVERSE LOAD

The following approximate solutions satisfying the geometric boundary conditions in terms of time-dependent amplitudes $\left(W_{mn}, \theta_{x_{mn}}, \theta_{y_{mn}} \right)$ are appropriate in the present problem:

$$\bar{w}(x,y,t) = \sum_{m=1}^{\infty}\sum_{n=1}^{\infty} W_{mn}(t)\sin\left(\frac{m\pi x}{a}\right)\sin\left(\frac{n\pi y}{b}\right)$$

$$\Phi_x(x,y,t) = \sum_{m=1}^{\infty}\sum_{n=1}^{\infty} \theta_{x_{mn}}(t)\cos\left(\frac{m\pi x}{a}\right)\sin\left(\frac{n\pi y}{b}\right) \tag{9.60}$$

$$\Phi_y(x,y,t) = \sum_{m=1}^{\infty}\sum_{n=1}^{\infty} \theta_{y_{mn}}(t)\sin\left(\frac{m\pi x}{a}\right)\cos\left(\frac{n\pi y}{b}\right)$$

Also, the solution is assumed for Airy's function (ϕ) which will satisfy the edge boundary condition and compatibility equation and is expressed as [42, 71, 72]:

$$\phi(x,y,t) = \frac{1}{2}N_{x0}y^2 + \frac{1}{2}N_{y0}x^2$$

$$+ \sum_{m=1}^{\infty}\sum_{n=1}^{\infty}\left(\varpi_{1_{mn}}(t)\cos\left(\frac{2m\pi x}{a}\right) + \varpi_{2_{mn}}(t)\cos\left(\frac{2n\pi y}{b}\right)\right) \tag{9.61}$$

The undetermined coefficients, $\varpi_{1_{mn}}(t)$ and $\varpi_{2_{mn}}(t)$, are determined by satisfying a compatibility equation. Thus, by substituting Equation 9.61 into Equation 9.53, the following unknowns are obtained:

$$\varpi_{1_{mn}}(t) = \frac{AW_{mn}(t)^2\beta^2}{32\alpha^2} \tag{9.62}$$

$$\varpi_{2_{mn}}(t) = \frac{AW_{mn}(t)^2\alpha^2}{32\beta^2} \tag{9.63}$$

where $\alpha = \dfrac{m\pi}{a}, \beta = \dfrac{n\pi}{b}$

The plate is harmonically excited with the mechanical and thermal load on the top surface for dynamic analysis, which is expressed as:

$$\{q_z(x,y),\Delta T(x,y)\} = \sum_{m=1}^{\infty}\sum_{n=1}^{\infty}\{Q_{mn},\Delta\bar{T}_{mn}\}\sin\left(\frac{m\pi x}{a}\right)\sin\left(\frac{n\pi y}{b}\right) \tag{9.64}$$

where $\{Q_{mn},\Delta\bar{T}_{mn}\} = \dfrac{4}{ab}\displaystyle\int_0^a\int_0^b \{q_z(x,y),\Delta T(x,y)\}\sin\left(\frac{m\pi x}{a}\right)\sin\left(\frac{n\pi y}{b}\right)dx\,dy$

For uniform mechanical and thermal load, $Q_{mn} = \dfrac{16Q_a}{mn\pi^2}, \bar{T}_{mn} = \dfrac{16\Delta T_i}{mn\pi^2}$.

9.5.2 Equivalent Axial Loads

The conditions of immovable boundaries are fulfilled by assuming the in-plane deflections δ_x and δ_y zero in the average sense, as expressed in Equation 9.65, in order to calculate equivalent compressive axial loads.

$$\delta_x = \int_\Omega \frac{\partial u}{\partial x}\,dxdy = 0; \quad \delta_y = \int_\Omega \frac{\partial v}{\partial y}\,dxdy = 0; \tag{9.65}$$

Also, from Equations 9.24–9.27 and 9.50–9.52 and simplifying, we get:

$$\frac{\partial \bar{u}}{\partial x} = \frac{1}{A}\left(N_{xx} - vN_{yy} - B\left(\varepsilon_b^{(1)}{}_{xx} + 9\varepsilon_s^{(1)}{}_{xx}\right) - F\varepsilon_s^{(1)}{}_{xx}\right) - \frac{1}{2}\left(\frac{\partial \bar{w}}{\partial x}\right)^2$$

$$\frac{\partial \bar{v}}{\partial y} = \frac{1}{A}\left(N_{yy} - vN_{xx} - B\left(\varepsilon_b^{(1)}{}_{yy} + 9\varepsilon_s^{(1)}{}_{yy}\right) - F\varepsilon_s^{(1)}{}_{yy}\right) - \frac{1}{2}\left(\frac{\partial \bar{w}}{\partial y}\right)^2 \tag{9.66}$$

Thus, equivalent compressive axial loads are found by substituting Equation 9.66 into Equation 9.65 and are expressed as:

$$N_{x0} = -\frac{\bar{A}_{mn}^T}{1-v} - \frac{4}{\pi^2 mn\left(1-v^2\right)}\left(\begin{array}{c}\left(B9+F\right)\left(\mu\theta_{x_{mn}}(t)+v\beta\theta_{y_{mn}}(t)\right)\\ -B\left(\mu^2 + v\beta^2\right)W_{mn}(t)\end{array}\right)$$

$$+\frac{A}{8\left(1-v^2\right)}\left(\mu^2 + v\beta^2\right)W_{mn}(t)^2$$

$$\tag{9.67}$$

$$N_{y0} = -\frac{\bar{A}_{mn}^T}{1-v} - \frac{4}{\pi^2 mn\left(1-v^2\right)}\left(\begin{array}{c}\left(B9+F\right)\left(v\mu\theta_{x_{mn}}(t)+\beta\theta_{y_{mn}}(t)\right)\\ -B\left(v\mu^2 + \beta^2\right)W_{mn}(t)\end{array}\right)$$

$$+\frac{A}{8\left(1-v^2\right)}\left(v\mu^2 + \beta^2\right)W_{mn}(t)^2$$

9.6 EQUATION OF MOTION

The equation of motion for dynamic analysis of a sandwich S-FGM plate is obtained using the Galerkin method. Thus, in the governing differential equation, Equations 9.57–9.59, the expressions for assumed approximate solutions (Equations 9.60–9.63) and transverse load (Equation 9.64) are substituted. In addition, for the Galerkin method to followed, the shape functions from Equation 9.60 are pre-multiplied by the corresponding equations in Equations 9.57–9.59 and integrated over the whole domain. This results in the following equation of motions:

$$\left\{ \begin{aligned} &M_{11}\ddot{W}_{mn}(t)+M_{12}\ddot{\theta}_{x_{mn}}(t)+M_{13}\ddot{\theta}_{y_{mn}}(t)+\left(\kappa_{11}-\psi_{11}\right)W_{mn}(t) \\ &+\kappa_{12}\theta_{x_{mn}}(t)+\kappa_{13}\theta_{y_{mn}}(t)+\kappa_{nl}W_{mn}(t)^{3} \end{aligned} \right\} = Q_{mn}+f_{1_{mn}} \quad (9.68)$$

$$M_{21}\ddot{W}_{mn}(t)+M_{22}\ddot{\theta}_{x_{mn}}(t)+M_{23}\ddot{\theta}_{y_{mn}}(t)+\kappa_{21}W_{mn}(t)$$
$$+\kappa_{22}\theta_{x_{mn}}(t)+\kappa_{23}\theta_{y_{mn}}(t)=f_{2_{mn}} \quad (9.69)$$

$$M_{31}\ddot{W}_{mn}(t)+M_{32}\ddot{\theta}_{x_{mn}}(t)+M_{33}\ddot{\theta}_{y_{mn}}(t)+\kappa_{31}W_{mn}(t)$$
$$+\kappa_{32}\theta_{x_{mn}}(t)+\kappa_{33}\theta_{y_{mn}}(t)=f_{3_{mn}} \quad (9.70)$$

The nonlinear Equation 9.68 is a duffing equation consisting of cubic nonlinearity and provides hard spring characteristics to the plate.

9.6.1 Forced Vibration Analysis

The second order nonlinear ordinary differential equation to analyze dynamic responses is obtained by substituting Equation 9.67 into Equation 9.68 which leads to the introduction of quadratic nonlinearity due to axial compressive loads resist in moving the edges in addition to cubic nonlinearity. The resulting equation obtained is solved using the Runge–Kutta fourth order method and is expressed as:

$$\left\{ \begin{aligned} &M_{11}\ddot{W}_{mn}(t)+M_{12}\ddot{\theta}_{x_{mn}}(t)+M_{13}\ddot{\theta}_{y_{mn}}(t)+\left(\kappa_{11}+\kappa^{T}\right)W_{mn}(t) \\ &+\kappa_{12}\theta_{x_{mn}}(t)+\kappa_{13}\theta_{y_{mn}}(t)+\tilde{\kappa}_{11}\theta_{x_{mn}}(t)W(t) \\ &+\tilde{\kappa}_{12}\theta_{y_{mn}}(t)W_{mn}(t)+\tilde{\kappa}_{13}W_{mn}(t)^{2}+\tilde{\kappa}W_{mn}(t)^{3} \end{aligned} \right\} = Q_{mn}+f_{1_{mn}} \quad (9.71)$$

$$M_{21}\ddot{W}_{mn}(t)+M_{22}\ddot{\theta}_{x_{mn}}(t)+M_{23}\ddot{\theta}_{y_{mn}}(t)+\kappa_{21}W_{mn}(t)$$
$$+\kappa_{22}\theta_{x_{mn}}(t)+\kappa_{23}\theta_{y_{mn}}(t)=f_{2_{mn}} \quad (9.72)$$

$$M_{31}\ddot{W}_{mn}(t)+M_{32}\ddot{\theta}_{x_{mn}}(t)+M_{33}\ddot{\theta}_{y_{mn}}(t)+\kappa_{31}W_{mn}(t)$$
$$+\kappa_{32}\theta_{x_{mn}}(t)+\kappa_{33}\theta_{y_{mn}}(t)=f_{3_{mn}} \quad (9.73)$$

where $Q_{mn}=\nabla_{mn}q(t)$, ∇_{mn} is governed by applied load such that load can be applied uniformly, bi-sinusoidally, act at a point, or varying uniformly. $q(t)=Q_{a}\sin(\omega t)$ expresses the applied loading intensity and type of variation of the load with time.

The expressions for stiffness and mass matrices are presented as:

$$M_{11} = \left(\frac{I_1^2}{I_0} - I_2 \right) \left(\alpha^2 + \beta^2 \right) - I_0 \quad M_{21} = \alpha \left(J_2 - \frac{I_1 J_1}{I_0} \right) \quad M_{31} = \beta \left(J_2 - \frac{I_1 J_1}{I_0} \right)$$

$$M_{12} = \alpha \left(J_2 - \frac{I_1 J_1}{I_0} \right) \quad M_{22} = \left(\frac{J_1^2}{I_0} - J_3 \right) \quad M_{32} = 0$$

$$M_{13} = \beta \left(J_2 - \frac{I_1 J_1}{I_0} \right) \quad M_{23} = 0 \quad M_{33} = \left(\frac{J_1^2}{I_0} - J_3 \right)$$

$$\kappa_{11} = \frac{1}{1-v^2} \left(\frac{B^2}{A} - D \right) \left(\mu^2 + \beta^2 \right)^2 - K_w - K_p \left(\mu^2 + \beta^2 \right)$$

$$\kappa_{12} = -\frac{\mu}{1-v^2} \left(\frac{BF}{A} + \frac{B^2 \vartheta}{A} - (D\vartheta + H) \right) \left(\mu^2 + \beta^2 \right)$$

$$\kappa_{13} = -\frac{\beta}{1-v^2} \left(\frac{BF}{A} + \frac{B^2 \vartheta}{A} - (D\vartheta + H) \right) \left(\mu^2 + \beta^2 \right)$$

$$\kappa_{nl} = -\frac{A}{16} \left(\mu^4 + \beta^4 \right), \ \psi_{11} = N_{x0} \mu^2 + N_{y0} \beta^2$$

$$\kappa_{21} = -\frac{\mu}{1-v^2} \left(\frac{BF}{A} + \frac{B^2 \vartheta}{A} - (D\vartheta + H) \right) \left(\mu^2 + \beta^2 \right)$$

$$\kappa_{22} = -\left[\begin{array}{c} \left(\frac{\mu^2}{1-v^2} + \frac{\beta^2}{2(1+v)} \right) \left(D\vartheta^2 + 2H\vartheta + J - \frac{B^2 \vartheta^2}{A} - \frac{F^2}{A} - 2\frac{B\vartheta F}{A} \right) \\ + \frac{1}{2(1+v)} \left(A\vartheta^2 + 2\vartheta K + L \right) \end{array} \right]$$

$$\kappa_{23} = -\frac{\mu\beta}{2(1-v)} \left(D\vartheta^2 + 2H\vartheta + J - \frac{B^2 \vartheta^2}{A} - \frac{F^2}{A} - 2\frac{B\vartheta F}{A} \right)$$

$$\kappa_{31} = -\frac{\beta}{1-v^2} \left(\frac{BF}{A} + \frac{B^2 \vartheta}{A} - (D\vartheta + H) \right) \left(\mu^2 + \beta^2 \right)$$

$$\kappa_{32} = -\frac{\mu\beta}{2(1-v)} \left(D\vartheta^2 + 2H\vartheta + J - \frac{B^2 \vartheta^2}{A} - \frac{F^2}{A} - 2\frac{B\vartheta F}{A} \right)$$

$$\kappa_{33} = -\left[\begin{array}{c} \left(\frac{\beta^2}{1-v^2} + \frac{\mu^2}{2(1+v)} \right) \left(D\vartheta^2 + 2H\vartheta + J - \frac{B^2 \vartheta^2}{A} - \frac{F^2}{A} - 2\frac{B\vartheta F}{A} \right) \\ + \frac{1}{2(1+v)} \left(A\vartheta^2 + 2\vartheta K + L \right) \end{array} \right]$$

$$f_{1_{mn}} = -\bar{B}^T \frac{\mu^2 + \beta^2}{1-\nu} \bar{T}_{mn},$$

$$f_{2_{mn}} = \left(\bar{F}^T + \bar{B}^T 9 \right) \frac{\mu}{1-\nu} \bar{T}_{mn},$$

$$f_{3_{mn}} = \left(\bar{F}^T + \bar{B}^T 9 \right) \frac{\beta}{1-\nu} \bar{T}_{mn}$$

$$\kappa^T = \frac{\bar{A}^T \left(\mu^2 + \beta^2 \right)}{1-\nu}, \quad \tilde{\kappa}_{11} = \frac{4\mu \left(B9 + F \right) \left(\mu^2 + \nu\beta^2 \right)}{\pi^2 mn \left(1 - \nu^2 \right)},$$

$$\tilde{\tilde{\kappa}}_{12} = \frac{4\beta \left(B9 + F \right) \left(\nu\mu^2 + \beta^2 \right)}{\pi^2 mn \left(1 - \nu^2 \right)},$$

$$\tilde{\tilde{\kappa}}_{13} = -\frac{4B \left(\mu^4 + 2\nu\mu^2\beta^2 + \beta^4 \right)}{\pi^2 mn \left(1 - \nu^2 \right)}, \quad \tilde{\tilde{\kappa}} = \left(\kappa_{nl} - \frac{A \left(\mu^4 + 2\nu\mu^2\beta^2 + \beta^4 \right)}{8 \left(1 - \nu^2 \right)} \right)$$

In the present formulation, uniformly distributed harmonic excitation is assumed, therefore, $\nabla_{mn} = 16/mn\pi^2$.

$$Q_{mn} = -\frac{16}{mn\pi^2} Q_a \sin\left(\omega t \right) \tag{9.74}$$

9.6.2   Free Vibration Analysis

The eigenvalue problem is derived from Equations 9.71–9.73 after eliminating the nonlinearity and applied load. Thus, the equation of motion for computing natural frequency is expressed as:

$$Q_{mn} = 0$$

$$\left| \kappa_{ij} - \omega^2 M_{ij} \right| = 0 \tag{9.75}$$

9.6.3   Static Analysis

The results for transverse displacement for static analysis are obtained using Equations 9.71–9.73 after eliminating the inertia element, in other words, the mass matrix. The equation is then expressed as:

$$\left\{ \begin{array}{l} \left(\kappa_{11} + \kappa^T \right) W_{mn}(t) + \kappa_{12}\theta_{x_{mn}}(t) + \kappa_{13}\theta_{y_{mn}}(t) + \tilde{\kappa}_{11}\theta_{x_{mn}}(t)W(t) \\[2mm] + \tilde{\kappa}_{12}\theta_{y_{mn}}(t)W_{mn}(t) + \tilde{\tilde{\kappa}}_{13}W_{mn}(t)^2 + \tilde{\tilde{\kappa}}W_{mn}(t)^3 \end{array} \right\} = Q_{mn} + f_{1_{mn}} \tag{9.76}$$

$$\kappa_{21}W_{mn}(t)+\kappa_{22}\theta_{x_{mn}}(t)+\kappa_{23}\theta_{y_{mn}}(t)=f_{2_{mn}} \tag{9.77}$$

$$\kappa_{31}W_{mn}(t)+\kappa_{32}\theta_{x_{mn}}(t)+\kappa_{33}\theta_{y_{mn}}(t)=f_{3_{mn}} \tag{9.78}$$

9.6.4 Relation between Linear Frequency, Nonlinear Frequency, and Load Amplitude with Displacement

The dependency of frequency on the amplitude of displacement and load occurs as a result of nonlinearity. The relationship is derived using one of the simplest methods known as the *harmonic balance method* (HBM) for which the equation of motion for forced vibration analysis (Equations 9.71–9.73) is simplified by neglecting the rotary inertia and rearranging the terms. Thus, the resulting simplified equation can be written as:

$$\left\{ \begin{aligned} & \left(\kappa_{11}+\kappa^{T}\right)W_{mn}(t)+\kappa_{12}\theta_{x_{mn}}(t)+\kappa_{13}\theta_{y_{mn}}(t) \\ & +\tilde{\kappa}_{11}\theta_{x_{mn}}(t)W_{mn}(t)+\tilde{\kappa}_{12}\theta_{y_{mn}}(t)W_{mn}(t) \\ & +\tilde{\kappa}_{13}W_{mn}(t)^{2}+\tilde{\tilde{\kappa}}W_{mn}(t)^{3}+Q_{mn} \end{aligned} \right\}=-M_{11}\ddot{W}_{mn}(t) \tag{9.79}$$

$$\kappa_{21}W_{mn}(t)+\kappa_{22}\theta_{x_{mn}}(t)+\kappa_{23}\theta_{y_{mn}}(t)=0 \tag{9.80}$$

$$\kappa_{31}W_{mn}(t)+\kappa_{32}\theta_{x_{mn}}(t)+\kappa_{33}\theta_{y_{mn}}(t)=0 \tag{9.81}$$

Obtaining $\theta_x(t)$ and $\theta_y(t)$ from Equation 9.80 and Equation 9.81, respectively, and substituting back into Equation 9.79, we get:

$$M_{11}\ddot{W}_{mn}(t)+\mho_{1}W_{mn}(t)+\mho_{2}W_{mn}(t)^{2}+\tilde{\tilde{\kappa}}W_{mn}(t)^{3}=-Q_{mn} \tag{9.82}$$

where

$$\mho_{1}=\kappa_{11}+\kappa^{T}+\kappa_{12}\left(\frac{\kappa_{23}\kappa_{31}-\kappa_{21}\kappa_{33}}{\kappa_{22}\kappa_{33}-\kappa_{23}\kappa_{32}}\right)+\kappa_{13}\left(\frac{\kappa_{21}\kappa_{32}-\kappa_{22}\kappa_{31}}{\kappa_{22}\kappa_{33}-\kappa_{23}\kappa_{32}}\right)$$

$$\mho_{2}=\tilde{\kappa}_{11}\left(\frac{\kappa_{23}\kappa_{31}-\kappa_{21}\kappa_{33}}{\kappa_{22}\kappa_{33}-\kappa_{23}\kappa_{32}}\right)+\tilde{\kappa}_{12}\left(\frac{\kappa_{21}\kappa_{32}-\kappa_{22}\kappa_{31}}{\kappa_{22}\kappa_{33}-\kappa_{23}\kappa_{32}}\right)+\tilde{\kappa}_{13}$$

The time-dependent transverse displacement, $W(t)$, in terms of maximum amplitude, w_{max}, of vibration can be written as:

$$W(t)=w_{max}\sin\left(\omega_{nl}t\right) \tag{9.83}$$

An algebraic equation relating linear frequency (ω_{mn}), nonlinear frequency (ω_{nl}), load (Q_a), and maximum displacement (w_{max}) is obtained using the Galerkin method. Thus, substituting Equation 9.74 and Equation 9.83 into Equation 9.82:

$$\omega_{nl}^2 = \omega_{mn}^2 \left(1 + \frac{8w_{max}}{3\pi} \frac{\mho_2}{\mho_1} + \frac{3w_{max}^2}{4} \frac{\tilde{\tilde{\kappa}}}{\mho_1} \right) - \frac{\nabla_{mn}}{M_{11}} \frac{Q_a}{w_{max}} \tag{9.84}$$

The relationship between load amplitude and linear to nonlinear frequency ratio is expressed as

$$\aleph^2 - \left(1 + \frac{8w_{max}}{3\pi} \frac{\mho_2}{\mho_1} + \frac{3w_{max}^2}{4} \frac{\tilde{\tilde{\kappa}}}{\mho_1} \right) = -\frac{Q_{max}}{\omega_{mn}^2 w_{max}} \tag{9.85}$$

where $\aleph = \dfrac{\omega_{nl}}{\omega_{mn}}$, $Q_{max} = \dfrac{\nabla_{mn}Q_a}{M_{11}}$

When no applied load is considered, i.e. $Q_{max} = 0$

$$\omega_{nl}^2 = \omega_{mn}^2 \left(1 + \frac{8w_{max}}{3\pi} \frac{\mho_2}{\mho_1} + \frac{3w_{max}^2}{4} \frac{\tilde{\tilde{\kappa}}}{\mho_1} \right) \tag{9.86}$$

Where $\omega_{mn} = \sqrt{-\dfrac{\mho_1}{M_{11}}}$

Equation 9.86 shows that nonlinear frequency is displacement dependent since, for Duffing nonlinearities, it is a well-known form of behavior.

9.7 VALIDATION AND CONVERGENCE STUDY

To formulate the governing equations for nonlinear dynamic analysis, a semi-analytical method is applied. Thus, it is important to validate the present mathematical formulation with the existing solution.

9.7.1 VALIDATION STUDY

To illustrate the preceding formulation, the solutions obtained here will be verified against the existing results of natural frequencies for FGM plates. No previous research on the nonlinear analysis of porous sandwich S-FGM plates has been performed under thermal environment, as stated previously. Therefore, the nonlinear frequency ratio is validated for ambient and thermal environment.

Thus, square $Si_3N_4/SUS304$ plates with thickness ratios $a/\hbar = 10$ and 20 are analyzed for computing nonlinear frequency ratio (ω_{nl}/ω_l). The present example demonstrates the validity of the formulation for nonlinear response using Equation 9.86.

TABLE 9.4

Validation of Nonlinear Frequency Ratio $\left(\omega_{nl}/\omega_l\right)$ for Different $(a/\hbar)$ and $(W_{max}/\hbar)$ of FGM Square Plate $(p = 1)$

	a/ℏ=10		a/ℏ=20	
$(w_{max}/\hbar)$	Ref. [73]	Present	Ref. [73]	Present
0.2	1.0063	1.0294	1.0053	1.0293
0.4	1.0654	1.0975	1.0617	1.0953
0.6	1.1707	1.1972	1.1627	1.1919
0.8	1.3115	1.3212	1.2989	1.3123
1.0	1.4789	1.4635	1.4614	1.4507

Table 9.4 contains the nonlinear frequency ratio of FGM plate for various values of $\left(w_{max}/\hbar\right)$. The material properties used in the present example are given as: $E_m = 207.7877 \times 10^9 \text{ N/m}^2$, $\rho_m = 8166 \text{ Kg/m}^3$ for 304 stainless steel (SUS304) and $E_c = 322.2715 \times 10^9 \text{ N/m}^2$, $\rho_c = 2370 \text{ Kg/m}^3$ for silicon nitride (Si$_3$N$_4$), $\nu = 0.28$. The present results are in good harmony with the existing results of Sundararajan et al. [73]. In contrast to the semi-analytical solution obtained in the present formulation for the nonlinear study of the FGM plate, they obtained a numerical solution. The discrepancy in results happens primarily because the current assessment is based on the HBM, while Sundararajan et al. [73] used the FEM in combination with the direct iteration technique. They also include rotary inertia which is ignored in the present case.

Afterward, the effect of the amplitude ratio $\left(w_{max}/\hbar\right)$ on the nonlinear frequency ratio $\left(\omega_{nl}/\omega_l\right)$ is perceived. The present formulation is validated with the results of Huang et al. [45] for a square Si$_3$N$_4$/SUS304 S-FGM simply supported plate. The material properties of the constituent material are listed in Table 9.3. Poisson's ratio is assumed to be constant and equal to 0.28. However, the results are in good harmony as shown in Table 9.5, but a small variance in the results may occur as a result of difference in techniques for obtaining the nonlinear frequency ratio, since, in the present study, the harmonic balance method is applied in contrast to the perturbation technique [45].

9.7.2 CONVERGENCE STUDY

To determine the number of terms (m, n) that should be included to converge the solution for nonlinear transverse displacement, a convergence study is performed. An Al$_2$O$_3$ homogeneous plate subjected to various magnitudes of uniform loading is considered. The present results obtained from a semi-analytical method for nonlinear transverse displacement are compared with the numerical solution of Azizian et al. [74]. The material properties used are given as: $E_m = 70 \times 10^9 \text{ N/m}^2$, $\rho_m = 2707 \text{ Kg/m}^3$ for Al

TABLE 9.5

Validation of Nonlinear Frequency Ratio $\left(\omega_{nl}/\omega_l\right)$ for Different $(w_{max}/\hbar)$ in a Thermal Environment (a/$\hbar$ = 8, a = 0.2, $\bar{K}_w = \bar{K}_p = 0$, $\Delta T = 100K$)

$(w_{max}/\hbar)$	Theory	Si_3N_4	$p=1$	SUS304
0	Huang et al. [45]	1	1	1
	Present	0.999	0.999	0.999
0.2	Huang et al. [45]	1.022	1.022	1.022
	Present	1.019	1.028	1.019
0.4	Huang et al. [45]	1.084	1.084	1.082
	Present	1.081	1.098	1.081
0.6	Huang et al. [45]	1.181	1.18	1.172
	Present	1.175	1.201	1.176
0.8	Huang et al. [45]	1.303	1.301	1.296
	Present	1.297	1.329	1.299
1	Huang et al. [45]	1.446	1.442	1.438
	Present	1.437	1.475	1.441

and $E_c = 380\times10$ N/m^2, $\rho_c = 3800$ Kg/m^3 for Al$_2$O$_3$, $\nu = 0.28$. In calculation of the transverse displacement, the converged solution is achieved by including more m and n terms. In addition, m and n are always odd integers, as for even entities, the expression for uniformly distributed load is zero. Although including 12 terms will give the solution closest to the numerical solution, as shown in Table 9.6, for the assessment of plates in dynamic analysis, it is computationally more costly at the expense of average error. Nevertheless, including the first four terms provides sufficient accuracy with an average error less than 0.22% which is acceptable for any computation.

9.8   RESULTS AND ANALYSIS

The present study is focused on nonlinear dynamic analysis of a plate under the influence of porosity defect and thermal environment. Thus, a porous sandwich plate is subjected to uniformly distributed non-dimensional load $\left(\bar{Q}_a = 50\right)$ at the top surface and bi-sinusoidal thermal load (ΔT=100 K or 600 K) across the thickness of the plate. Table 9.3 contains the material property considered for the analysis. The effects of porosity distributions and temperature, with various geometric parameters, on a sandwich S-FGM plate have been done using *"Time displacement responses"*, *"Phase-plane plots"*, and *"Poincaré maps"*. Unless otherwise defined, in the succeeding paragraphs, it is presumed that: $a/\hbar = 10$, $p = 4$, $\bar{K}_w = \bar{K}_p = 0$, $b/a = 1$, $e = 0.5$.

Material: ZrO$_2$/SUS304.

The Pasternak foundation in the non-dimensional form is expressed as:

$$\bar{K}_w = \frac{K_w a^4}{D}, \bar{K}_p = \frac{K_p a^2}{D}, D = \frac{E_m \hbar^3}{12\left(1-\nu^2\right)}$$

TABLE 9.6

Non-Dimensional Nonlinear Transverse Displacement $\left(\tilde{w} = 200\bar{w}\left(a/2, b/2\right)/a\right)$ for Al_2O_3 Homogeneous Plate under Non-Dimensional Uniform Load $\left(\bar{Q}_a = 12Q_a a^4\left(1-v^2\right)/E_c\,\hbar^4\right)$ for Convergence Study $\left(a/\hbar = 20\right)$

Theory	Terms	$\bar{Q}_a = 10$ (#)	$\bar{Q}_a = 50$ (#)	$\bar{Q}_a = 75$ (#)	$\bar{Q}_a = 100$ (#)
Present	(1,1)	$0.4209^{(2.538)}$	$2.0025^{(2.677)}$	$2.8541^{(2.813)}$	$3.5977^{(2.974)}$
	(1,1) (3,1)	$0.4150^{(1.101)}$	$1.9728^{(1.154)}$	$2.8096^{(1.210)}$	$3.5383^{(1.274)}$
	(1,1) (3,1) (1,3) (3,3)	$\mathbf{0.4097^{(0.190)}}$	$\mathbf{1.9463^{(0.205)}}$	$\mathbf{2.7699^{(0.220)}}$	$\mathbf{3.4854^{(0.240)}}$
	(1,1) (3,1) (5,1) (1,3) (3,3) (5,3)	$0.4102^{(0.068)}$	$1.9486^{(0.087)}$	$2.7734^{(0.094)}$	$3.49^{(0.109)}$
	(1,1) (3,1) (5,1) (1,3) (3,3) (5,3) (1,5) (3,5) (5,5)	$0.4107^{(0.054)}$	$1.9511^{(0.041)}$	$2.7771^{(0.040)}$	$3.495^{(0.034)}$
	(1,1) (3,1) (5,1) (7,1) (1,3) (3,3) (5,3) (7,3) (1,5) (3,5) (5,5) (7,5)	$0.4106^{(0.029)}$	$1.9506^{(0.015)}$	$2.7763^{(0.011)}$	$3.4939^{(0.003)}$
Azizian et al. [74]	$(3,4)^*$	0.41048	1.9503	2.776	3.4938

$^{\#}$Error (%), $^{*}(n_s, n_0)$, n_s=number of strips, n_0= number of terms used in finite strip method.

9.8.1 Effect of Span-to-Thickness Ratio

In this section, very thick $\left(a/\hbar = 5\right)$ and thick $\left(a/\hbar = 10\right)$ (1-1-1) plate is analyzed and studied for predicting the dynamic behavior with $P-1$ and $P-3$ porosity distribution as shown in Figure 9.6 and Figure 9.7, respectively, under thermal environment. The modal frequencies for very thick $\left(a/\hbar = 5\right)$ and thick $\left(a/\hbar = 10\right)$ sandwich S-FGM plate (1-1-1) with $P-1/P-3$ porosity distribution are 7218/6690 rad/s (1148/1065 Hz) and 4021/3655 rad/s (640/582 Hz) respectively at ΔT= 100 K, and 6248/5677 rad/s (994/904 Hz) and 2976/2344 rad/s (474/373 Hz) respectively at ΔT=600 K. It has been observed that the system dynamic response is sensitive to $P-1$ and $P-3$ porosity as with $P-1$ porosity, the system nature is quasi-periodic with a strange attractor, which has multi-discrete points/lobes as shown in Figure 9.6. While for $P-3$ porosity distribution, the system is multi-periodic in nature with a weak attractor.

In addition, for very thick $\left(a/\hbar = 5\right)$ and thick $\left(a/\hbar = 10\right)$ (2-2-1) plate, the modal frequencies are 6847 rad/s (1090 Hz) and 3737 rad/s (595 Hz), respectively, at ΔT= 100 K, and 2778 rad/s (442 Hz) and 2341 rad/s (372 Hz), respectively, at ΔT=600 K as shown in Figure 9.8 with $P-3$ porosity distribution. The nature of the system is quasi-periodic with a very weak attractor. In consistent to the previous discussion on perfect plate, result trends are similar for the present case as well. Moreover, it is observed that a sandwich plate with $P-1$ porosity distribution results in higher modal frequency as compared to $P-3$ porosity distribution regardless of thermal environment and plate configurations.

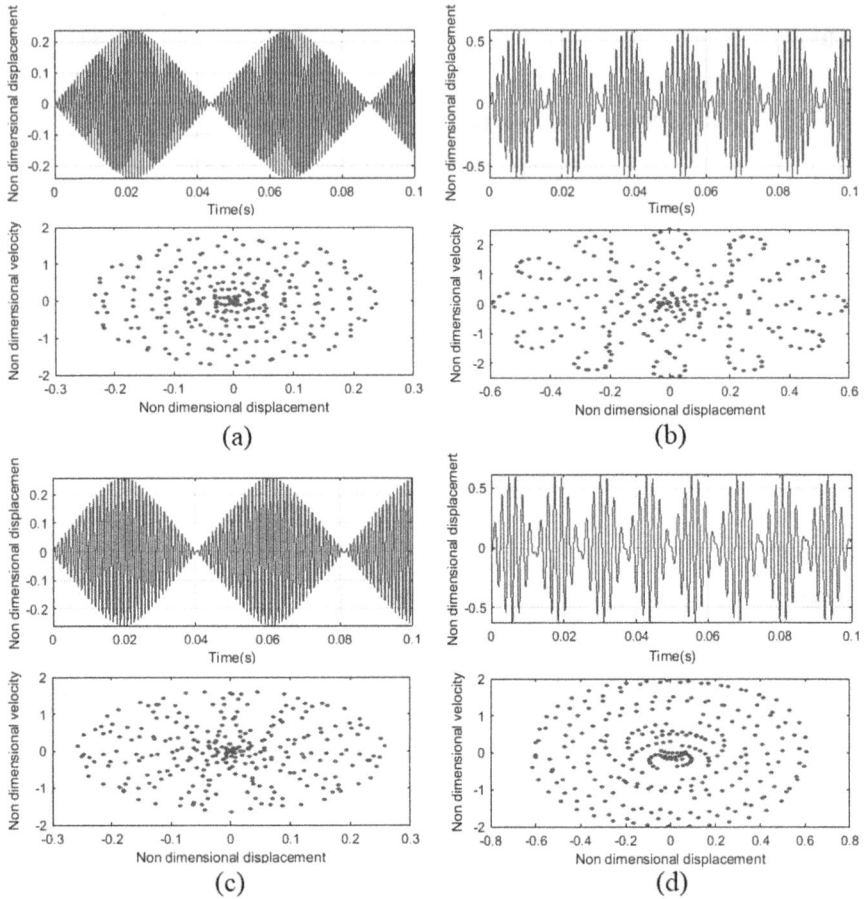

FIGURE 9.6 Nonlinear dynamic responses of (1-1-1) porous S-FGM sandwich plate at different thermal loads with P − 1 porosity distribution. (a) $a/\hbar = 5$, $\Delta T = 100$ K, (b) $a/\hbar = 10$, $\Delta T = 100$ K, (c) $a/\hbar = 5$, $\Delta T = 600$ K, (d) $a/\hbar = 10$, $\Delta T = 600$ K.

9.8.2 Effect of Aspect Ratio

In this section, $(b/a = 0.5)$ and $(b/a = 2)$ (2-1-2) plate is analyzed and studied for predicting the dynamic behavior as shown in Figure 9.9. The modal frequencies for $(b/a = 0.5)$ and $(b/a = 2)$ plate with $P - 3$ porosity distribution are 8444 rad/s (1344 Hz) and 2210 rad/s (352 Hz), respectively, at $\Delta T = 100$ K and 6987 rad/s (1112 Hz) and 1063 rad/s (169 Hz), respectively, at $\Delta T = 600$ K. The nature of the system is transforming from quasi-periodic to the onset of chaos with a rise in temperature for $P - 3$ porosity.

In addition, $(b/a = 0.5)$ and $(b/a = 2)$ (2-1-1) plate is analyzed and studied for predicting the dynamic behavior as depicted in Figure 9.10. The modal frequencies for $(b/a = 0.5)$ and $(b/a = 2)$ plate with $P - 3$ porosity distribution are 8584 rad/s (1366 Hz) and 2242 rad/s (357 Hz), respectively, at $\Delta T = 100$ K and 7090 rad/s (1128 Hz) and

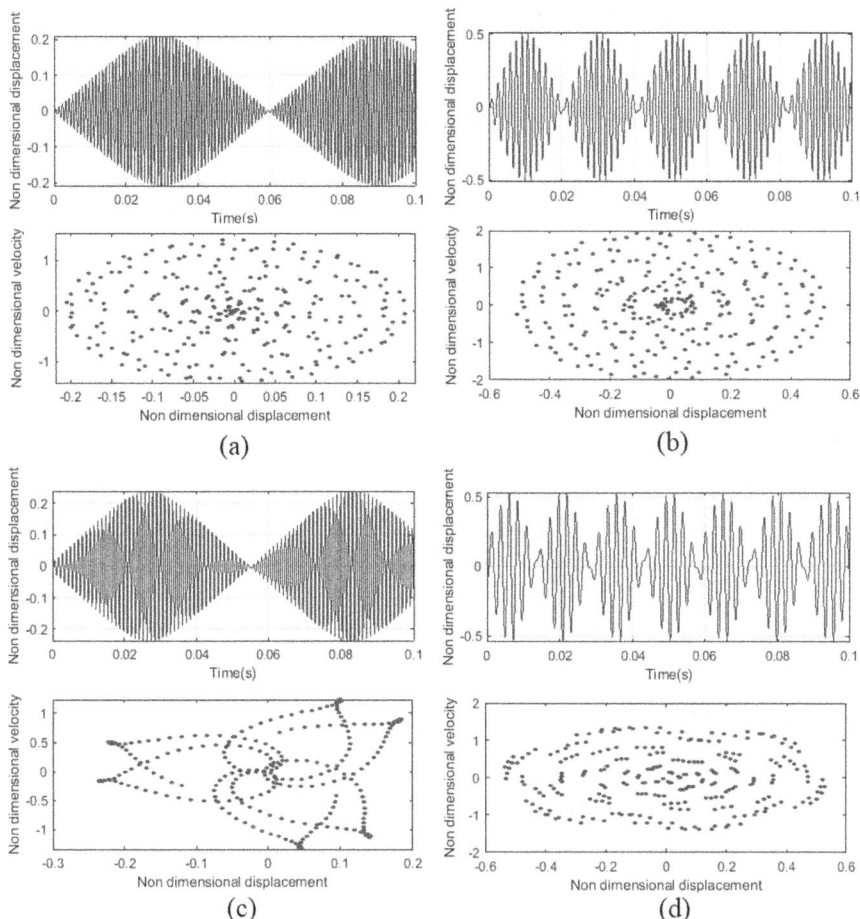

FIGURE 9.7 Nonlinear dynamic responses of (1-1-1) porous S-FGM sandwich plate at different thermal loads with P – 3 porosity distribution. (a) $a/\hbar = 5$, $\Delta T = 100$ K, (b) $a/\hbar = 10$, $\Delta T = 100$ K, (c) $a/\hbar = 5$, $\Delta T = 600$ K, (d) $a/\hbar = 10$, $\Delta T = 600$ K.

1064 rad/s (169 Hz), respectively, at $\Delta T = 600$ K. The system response shows the multi-periodic nature at $\Delta T = 100$ K, while the system response becomes chaotic with an increase in temperature difference. It is observed that asymmetric plate configuration (2-1-1) has higher modal frequency than symmetric plate (2-1-2) configuration irrespective of the thermal environment and aspect factor. Also, modal frequency decreases with increases in temperature and aspect factor. This result is anticipated because of the reduction in rigidity of the plate.

9.8.3 Effect of Volume Fraction Exponent

The variation in volume fraction exponent significantly affects the material property and hence its stiffness and density. This significant variation, consecutively, greatly

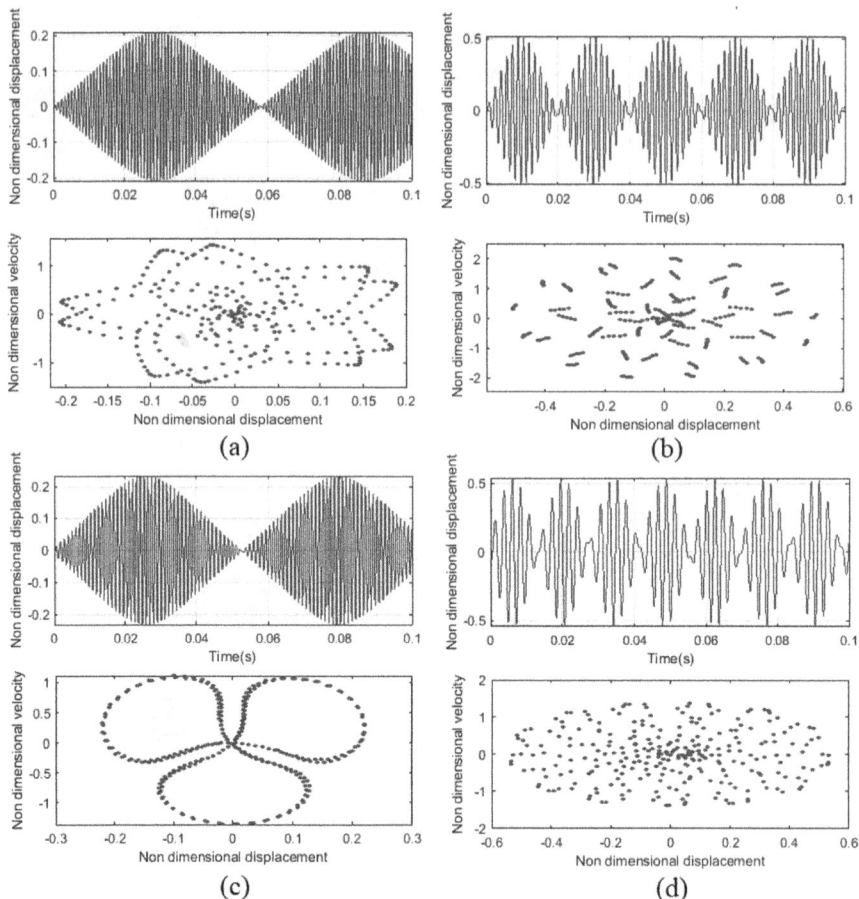

FIGURE 9.8 Nonlinear dynamic responses of (2-2-1) porous S-FGM sandwich plate at different thermal loads with P − 3 porosity distribution. (a) $a/\hbar = 5$, $\Delta T = 100$ K, (b) $a/\hbar = 10$, $\Delta T = 100$ K, (c) $a/\hbar = 5$, $\Delta T = 600$ K, (d) $a/\hbar = 10$, $\Delta T = 600$ K.

affects the dynamic behavior of the FGM plate as shown in Figure 9.11 and Figure 9.12 at $\Delta T = 100$ K and $\Delta T = 600$ K, respectively. The modal frequencies obtained for ZrO_2 and ($p=2$) sandwich plate are found to be highest for a non-symmetric (2-2-1) plate configuration with 3635 rad/s (579 Hz) and 3740 rad/s (595 Hz), respectively, and lowest for a symmetric (2-1-2) plate configuration with 3408 rad/s (542 Hz) and 3533 rad/s (562 Hz), respectively, at $\Delta T = 100$ K. However, at $\Delta T = 600$ K, the modal frequencies obtained for ZrO_2 and ($p=2$) sandwich plate are found to be highest for a symmetric (2-1-2) plate configuration with 1554 rad/s (247 Hz) and 2242 rad/s (357 Hz), respectively, and lowest for a non-symmetric (2-2-1) plate configuration with 1425 rad/s (227 Hz) and 2147 rad/s (342 Hz), respectively. The system responses are chaotic with a weak attractor (Figure 9.11) and a strong attractor (Figure 8.16).

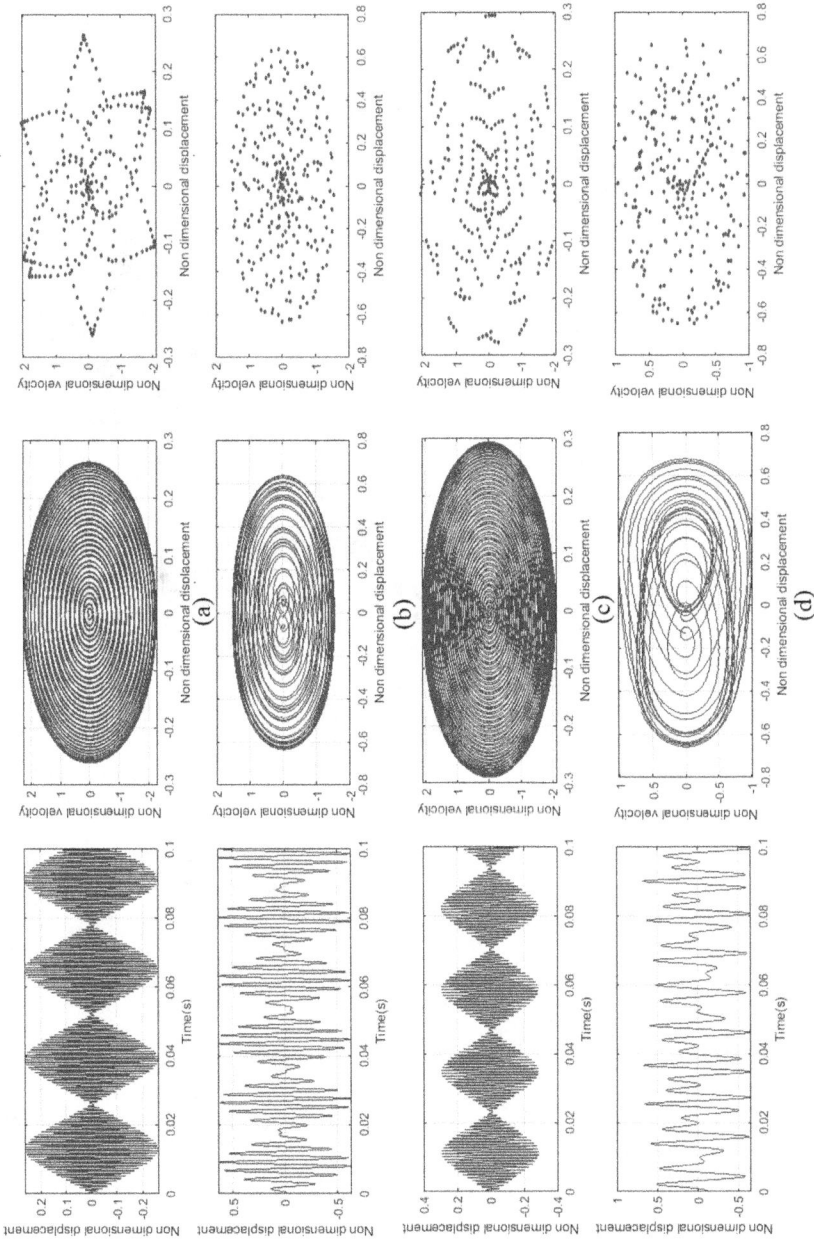

FIGURE 9.9 Nonlinear dynamic responses of (2-1-2) porous S-FGM sandwich plate at different thermal loads with P – 3 porosity distribution. (a) b/a=0.5, ΔT=100 K, (b) b/a=2, ΔT=100 K, (c) b/a=0.5, ΔT=600 K, (d) b/a=2, ΔT=600 K.

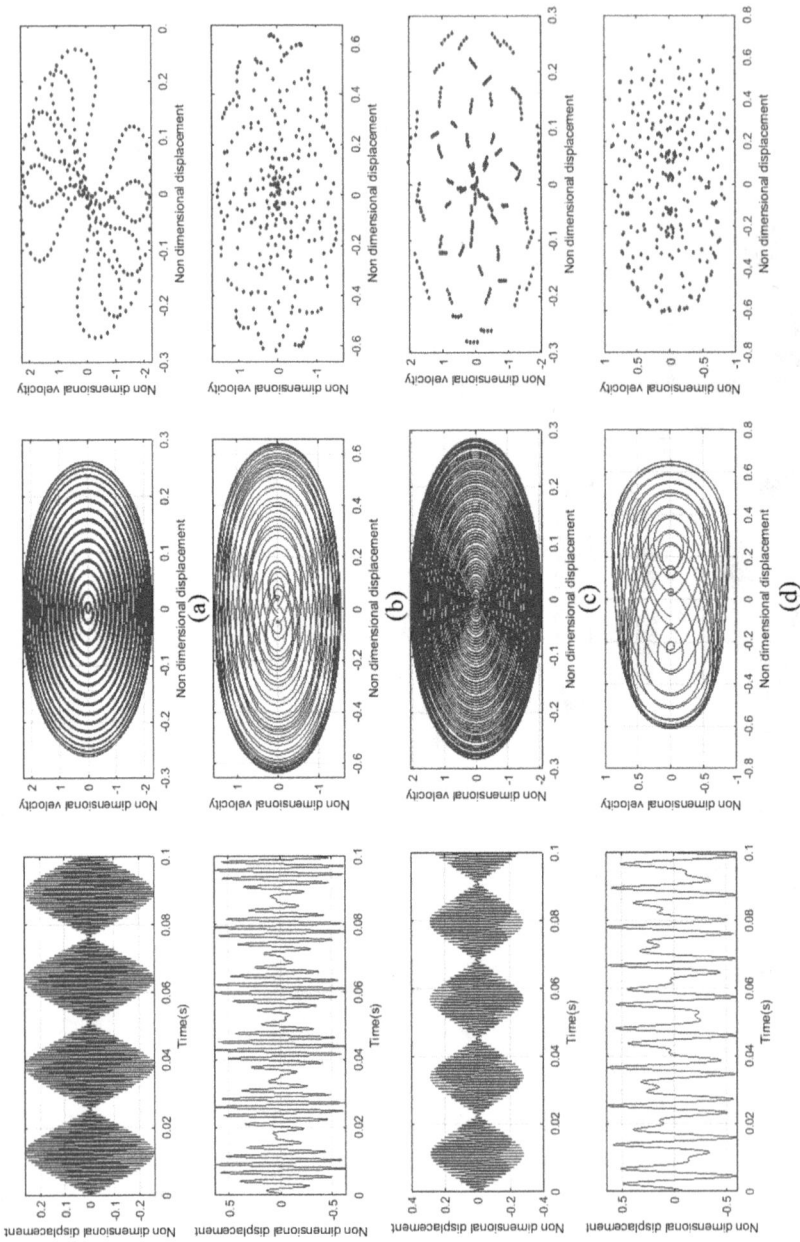

FIGURE 9.10 Nonlinear dynamic responses of (2-1-1) porous S-FGM sandwich plate at different thermal loads with P − 3 porosity distribution. (a) $b/a=0.5$, $\Delta T=100$ K, (b) $b/a=2$, $\Delta T=100$ K, (c) $b/a=0.5$, $\Delta T=600$ K, (d) $b/a=2$, $\Delta T=600$ K.

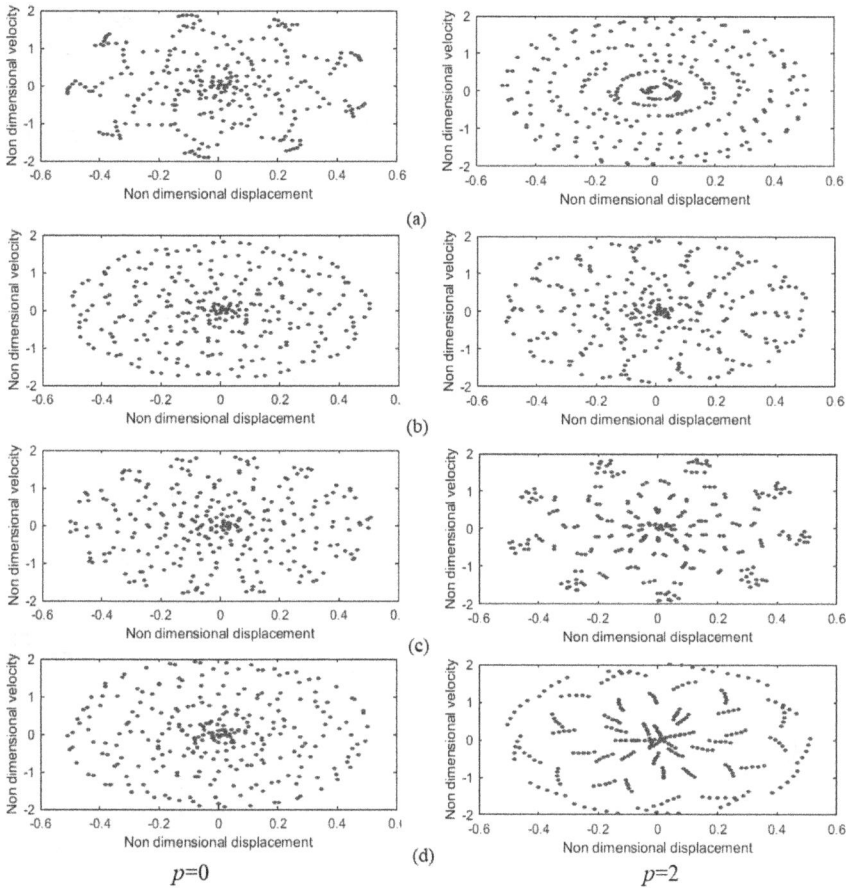

FIGURE 9.11 Nonlinear dynamic responses at $\Delta T = 100$ K with P – 3 porosity distribution for (a) 1-1-1, (b) 2-1-2, (c) 2-1-1, (d) 2-2-1 porous S-FGM sandwich plate.

The increases in modal frequency with the rise in "p" are paradoxical, since an increase in "p" results in an increase in the metallic content of the plate and lessens the stiffness of the plate. However, in the present study, SUS304 (metal) has a higher Young's modulus in comparison to ZrO_2 (ceramic) and therefore the stiffness of the plate increases with an increase in volume fraction of SUS304 irrespective of the temperature differences.

The above responses clearly show the effect of porosity on system dynamic response as with $P – 3$ the porosity system becomes more chaotic.

9.8.4 Effect of Elastic Foundation Parameters

Figure 9.13 and Figure 9.14 depict the effects of a Winkler–Pasternak foundation on the dynamic behavior, respectively, for (2-1-2) and (2-2-1) S-FGM sandwich

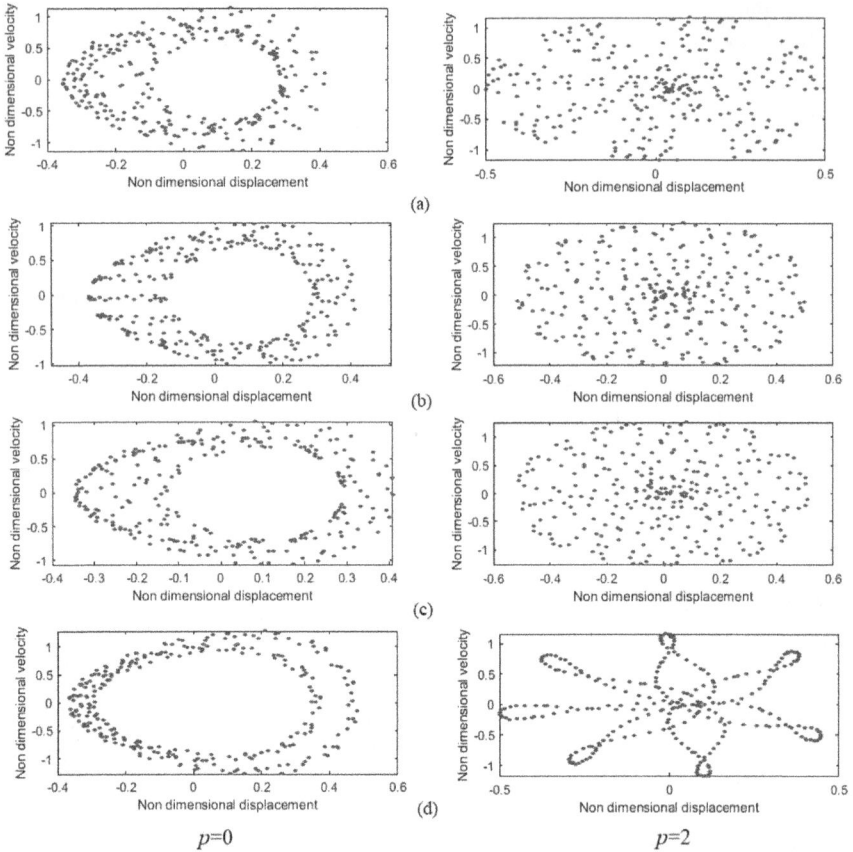

FIGURE 9.12 Nonlinear dynamic responses at $\Delta T = 600$ K with P – 3 porosity distribution for (a) 1-1-1, (b) 2-1-2, (c) 2-1-1, (d) 2-2-1 porous S-FGM sandwich plate.

plates. The modal frequencies for a plate resting only on a Winkler foundation $\left(\bar{K}_w = 100, \bar{K}_p = 0\right)$ and a Pasternak foundation $\left(\bar{K}_w = 0, \bar{K}_p = 100\right)$ are highest for P – 1 porosity distribution, and are given as 4733 rad/s (318.8 Hz) and 13,232 rad/s (738.2 Hz), respectively, for a 2-1-2 plate configuration and 5280 rad/s (318.8 Hz) and 15,017 rad/s (738.2 Hz), respectively, for a 2-2-1 plate configuration at $\Delta T = 100$ K. The responses illustrate the chaotic behavior with a strong attractor for P – 3 porosity, while for a P – 1 system, its nature is chaotic but with a weak attractor as shown in Figure 9.13 and Figure 9.14.

In addition, analysis is also performed at $\Delta T = 600$ K for different configurations and the plate is presumed to be resting only on a Pasternak foundation $\left(\bar{K}_w = 0, \bar{K}_p = 100\right)$. The plates of different configurations are excited at modal frequencies of 11,043 rad/s, 10,718 rad/s, 10,474 rad/s, and 10,292 rad/s, respectively, for 2-2-1, 1-1-1, 2-1-1, and 2-1-2 plates as shown in Figure 9.15. The system responses

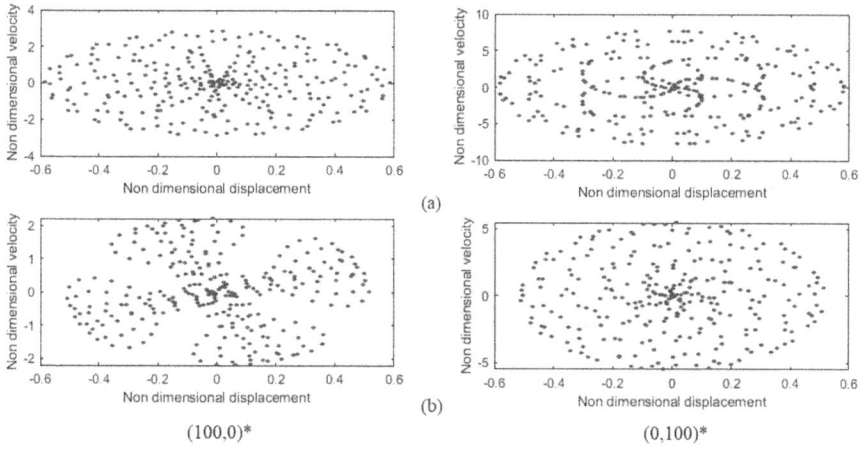

FIGURE 9.13 Nonlinear dynamic responses at $\Delta T = 100$ K for porous S-FGM (2-1-2) sandwich plate with (a) P – 1 and (b) P – 3 porosity distribution. $^*\left(\bar{K}_w, \bar{K}_p\right)$.

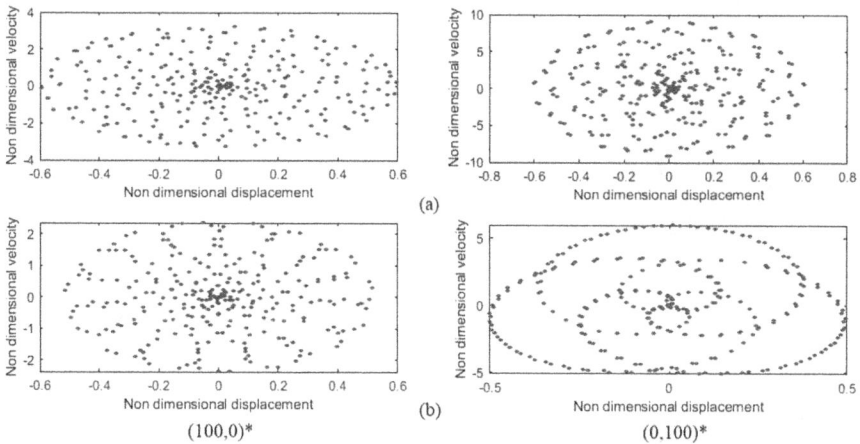

FIGURE 9.14 Nonlinear dynamic responses at $\Delta T = 100$ K for porous S-FGM (2-2-1) sandwich plate with (a) P – 1 and (b) P – 3 porosity distribution. $^*\left(\bar{K}_w, \bar{K}_p\right)$.

show the quasi-periodicity with different plate configurations; the attractor is strong for symmetric plate configuration.

9.8.5 Effect of Porosity Coefficient

Figure 9.16 and Figure 9.17 depict the effect of the porosity coefficient (e) on nonlinear dynamic responses for symmetric and asymmetric sandwich S-FGM plate, respectively. It is observed that with an increase in porosity coefficient, modal

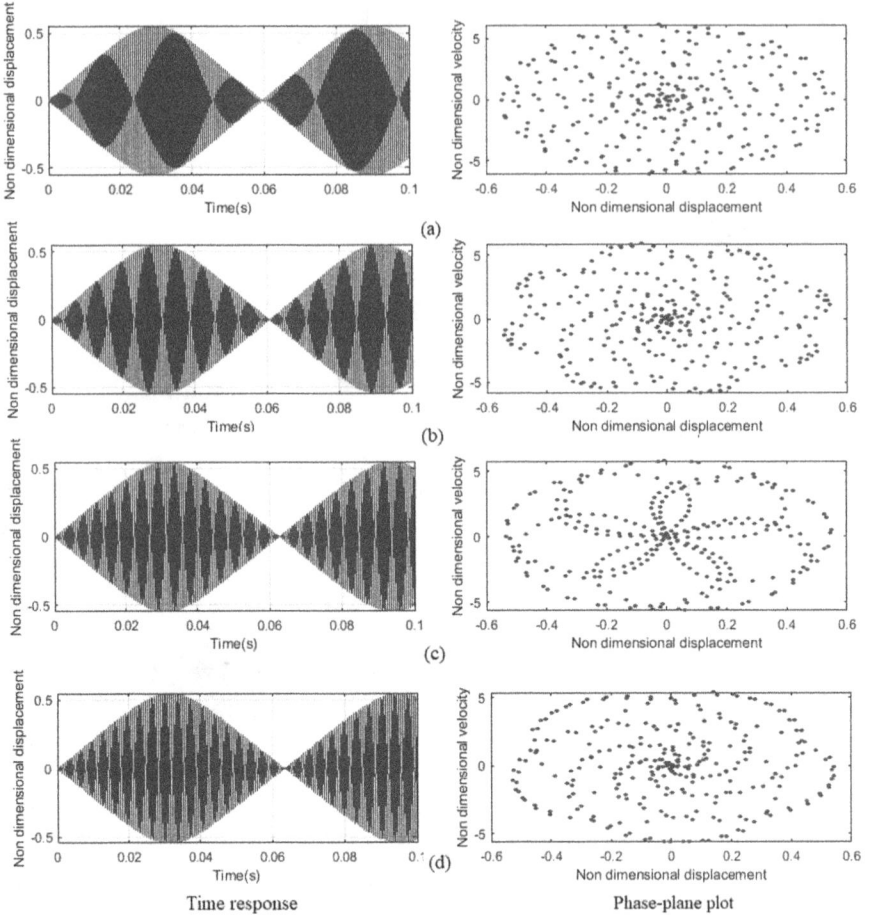

Time response   Phase-plane plot

FIGURE 9.15 Nonlinear dynamic responses at $\Delta T = 600$ K for (a) 2-2-1, (b) 1-1-1, (c) 2-1-1, (d) 2-1-2 porous S-FGM sandwich plate with P − 3 porosity distribution. $\left(\overline{K}_w = 0,\ \overline{K}_p = 100 \right)$.

frequency increases unexpectedly. This is due to the fact that increase in volume fraction of the pores will result in significant reduction in mass in comparison to the stiffness. This will sequentially increases the frequency irrespective of plate configuration and thermal environment. For symmetric plate configuration, the highest modal frequency is 3694 rad/s (588 Hz) when $e = 0.6$ and the lowest is 1551 rad/s (247 Hz) when $e = 0$ for a 1-1-1 plate configuration. Moreover, for asymmetric plate configuration, the highest modal frequency is 3772 rad/s (600 Hz) when $e = 0.6$ and the lowest is 1470 rad/s (234 Hz) when $e = 0$ for a 2-2-1 plate configuration. The plate behavior transforms from chaotic to multi-periodic as the porosity coefficient increases.

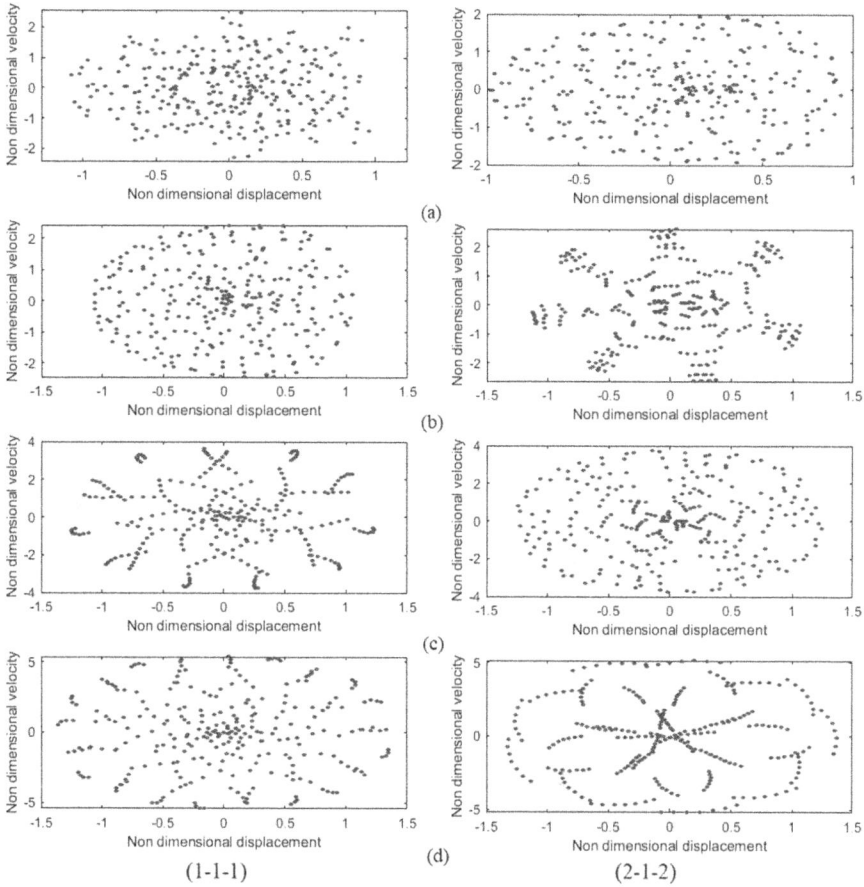

FIGURE 9.16   Nonlinear dynamic responses at $\Delta T = 600$ K for symmetric porous plate (a) $e = 0$, (b) $e = 0.1$, (c) $e = 0.3$, (d) $e = 0.6$ with P – 3 porosity distribution. $\left(a \, / \, \hbar = 20, \ \bar{K}_w = \bar{K}_p = 50 \right)$.

9.9   CONCLUSIONS

In this study, the sandwich plate with a homogeneous core is analyzed for the nonlinearity effect due to von Karman strain. The sandwich plate is subjected to harmonic load and presumed to be built on the Pasternak foundation. The temperature distribution varies linearly or nonlinearly across the thickness whereas the material properties of functionally graded face sheets are varied per sigmoid law (S-FGM). To solve geometric nonlinearity and implementing the different edge conditions, Airy's stress function and the Galerkin method have been implemented. The present formulation is validated with the results of available theory for different geometric parameters, elastic foundation, and plate configuration by considering a few numerical examples.

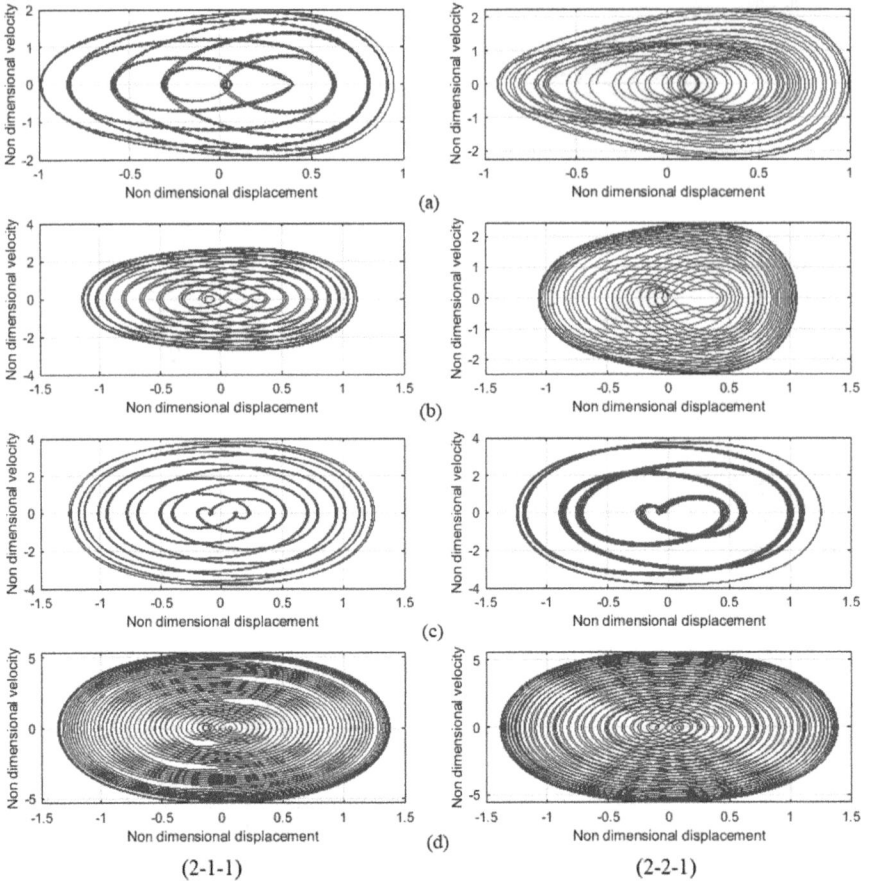

FIGURE 9.17 Nonlinear dynamic responses at $\Delta T=600$ K for non-symmetric porous plate (a) $e=0$, (b) $e=0.1$, (c) $e=0.3$, (d) $e=0.6$ with P – 3 porosity distribution. $\left(a/\hbar=20,\ \bar{K}_w=\bar{K}_p=50\right)$.

The present study assists researchers in actively controlling the porous plates and shells under thermodynamic loading. The subsequent points are significant:

1. The present formulation is validated with the results of available theory for different geometric parameters, elastic foundation, and plate configuration. The results are found to be very accurate.
2. The increase in $\left(a/\hbar\right)$ and moving from $P - 1$ porosity to $P - 3$ porosity results in a significant decrease in fundamental frequency.
3. It is observed that for $P - 1$ porosity, the system nature is quasi-periodic while for $P - 3$ porosity the distribution system nature is multi-periodic with a weak attractor with regard to $\left(a/\hbar\right)$.

4. As (b/a) increases, the plate behavior transforms from quasi-periodic to the onset of chaos with a rise in temperature for $P - 3$ porosity.
5. It is observed that an asymmetric plate configuration (2-1-1) has a higher modal frequency than a symmetric plate (2-1-2) configuration irrespective of the thermal environment and aspect factor.
6. The increases in modal frequency with the rise in volume fraction exponent are paradoxical in the present study.
7. The plate behavior shows quasi-periodicity with different plate configurations, for which the attractor is strong for symmetric plate configuration with regard to the presence of a Pasternak foundation.
8. As the porosity coefficient increases, the system nature transforms from chaotic to multi-periodic for a symmetric plate configuration.

REFERENCES

1. Koizumi M, Niino M (1995) Overview of FGM research in Japan. *MRS Bull* 20:19–21. https://doi.org/10.1557/S0883769400048867
2. Mortensen A, Suresh S (1995) Functionally graded metals and metal-ceramic composites. 1. Processing. *Int Mater Rev* 40:239–265. https://doi.org/10.1179/imr.1995.40.6.239
3. Zenkour AM (2005) A comprehensive analysis of functionally graded sandwich plates: Part 1-Deflection and stresses. *Int J Solids Struct* 42:5224–5242. https://doi.org/10.1016/j.ijsolstr.2005.02.015
4. Zenkour AM (2013) Bending analysis of functionally graded sandwich plates using a simple four-unknown shear and normal deformations theory. *J Sandw Struct Mater* 15:629-656. https://doi.org/10.1177/1099636213498886
5. Zenkour AM (2018) A quasi-3D refined theory for functionally graded single-layered and sandwich plates with porosities. *Compos Struct* 201:38–48. https://doi.org/10.1016/j.compstruct.2018.05.147
6. Tounsi A, Houari MSA, Benyoucef S, Adda Bedia EA (2013) A refined trigonometric shear deformation theory for thermoelastic bending of functionally graded sandwich plates. *Aerosp Sci Technol* 24:209–220. https://doi.org/10.1016/j.ast.2011.11.009
7. Taibi FZ, Benyoucef S, Tounsi A, et al. (2015) A simple shear deformation theory for thermo-mechanical behaviour of functionally graded sandwich plates on elastic foundations. *J Sandw Struct Mater* 17:99–129.https://doi.org/10.1177/1099636214554904
8. Bessaim A, Houari MSA, Tounsi A, et al. (2013) A new higher-order shear and normal deformation theory for the static and free vibration analysis of sandwich plates with functionally graded isotropic face sheets. *J Sandw Struct Mater* 15:671–703. https://doi.org/10.1177/1099636213498888
9. Di Sciuva M, Sorrenti M (2019) Bending, free vibration and buckling of functionally graded carbon nanotube-reinforced sandwich plates, using the extended refined zigzag theory. *Compos Struct* 227:111324. https://doi.org/10.1016/j.compstruct.2019.111324
10. Di Sciuva M, Sorrenti M (2019) Bending and free vibration analysis of functionally graded sandwich plates: An assessment of the refined zigzag theory. *J Sandw Struct Mater* 109963621984397. https://doi.org/10.1177/1099636219843970
11. Demirhan P, Taşkın V (2018) Static analysis of simply supported functionally graded sandwich plates by using four variable plate theory. *Tek Dergi.* 24:434–448. https://doi.org/10.18400/tekderg.396672

12. Demirhan PA, Taskin V (2018) Bending and free vibration analysis of Levy-type porous functionally graded plate using state space approach. *Compos Part B Eng.* 160:661–676. https://doi.org/10.1016/j.compositesb.2018.12.020

13. Sobhy M (2015) Thermoelastic response of FGM Plates with temperature-dependent properties resting on variable elastic foundations. *Int J Appl Mech* 07:1550082. https://doi.org/10.1142/s1758825115500829

14. Singh SJ, Harsha SP (2020) Thermo-mechanical analysis of porous sandwich S-FGM plate for different boundary conditions using Galerkin Vlasov's method: A semi-analytical approach. *Thin-Walled Structures* 150:106668. https://doi.org/10.1016/j.tws.2020.106668

15. Garg A, Belarbi MO, Chalak HD, Chakrabarti A (2021) A review of the analysis of sandwich FGM structures. *Compos. Struct.*258:113527.

16. Zenkour AM (2005) A comprehensive analysis of functionally graded sandwich plates: Part 2-Buckling and free vibration. *Int J Solids Struct* 42:5243–5258. https://doi.org/10.1016/j.ijsolstr.2005.02.016

17. Kurpa LV, Shmatko TV. (2020) Buckling and free vibration analysis of functionally graded sandwich plates and shallow shells by the Ritz method and the R-functions theory. *Proc Inst Mech Eng C* https://doi.org/10.1177/0954406220936304

18. Hadji L, Atmane HA, Tounsi A, et al (2011) Free vibration of functionally graded sandwich plates using four-variable refined plate theory. *Appl Math Mech* 32:925–942. https://doi.org/10.1007/s10483-011-1470-9

19. Singh SJ, Harsha SP (2019) Exact solution for free vibration and buckling of sandwich S-FGM plates on Pasternak elastic foundation with various boundary conditions. *Int J Struct Stab Dyn* 19:S0219455419500287. https://doi.org/10.1142/S0219455419500287

20. Sobhy M, Chakraverty S, Pradhan KK, et al (2014) Buckling and free vibration of exponentially graded sandwich plates resting on elastic foundations under various boundary conditions. *Int J Solids Struct* 43:76–87. https://doi.org/10.1177/1077546310370691

21. Meziane MAA, Abdelaziz HH, Tounsi A (2014) An efficient and simple refined theory for buckling and free vibration of exponentially graded sandwich plates under various boundary conditions. *J Sandw Struct Mater* 16:293–318. https://doi.org/10.1177/1099636214526852

22. Singh SJ, Harsha SP (2020) Analysis of porosity effect on free vibration and buckling responses for sandwich sigmoid function based functionally graded material plate resting on Pasternak foundation using Galerkin Vlasov's method. *J Sandw Struct Mater* https://doi.org/10.1177/1099636220904340

23. Wattanasakulpong N, Prusty GB, Kelly DW (2013) Free and forced vibration analysis using improved third-order shear deformation theory for functionally graded plates under high temperature loading. *J Sandw Struct Mater* 15:583–606. https://doi.org/10.1177/1099636213495751

24. El Meiche N, Tounsi A, Ziane N, et al (2011) A new hyperbolic shear deformation theory for buckling and vibration of functionally graded sandwich plate. *Int J Mech Sci* 53:237–247. https://doi.org/10.1016/j.ijmecsci.2011.01.004

25. Thai HT, Nguyen TK, Vo TP, Lee J (2014) Analysis of functionally graded sandwich plates using a new first-order shear deformation theory. *Eur J Mech A Solids* 45:211–225. https://doi.org/10.1016/j.euromechsol.2013.12.008

26. Nguyen VH, Nguyen TK, Thai HT, Vo TP (2014) A new inverse trigonometric shear deformation theory for isotropic and functionally graded sandwich plates. *Compos B Eng* 66:233–246. https://doi.org/10.1016/j.compositesb.2014.05.012

27. Xiang S, Kang G wen, Yang M sui, Zhao Y (2013) Natural frequencies of sandwich plate with functionally graded face and homogeneous core. *Compos Struct* 96:226–231. https://doi.org/10.1016/j.compstruct.2012.09.003

28. Neves AMA, Ferreira AJM, Carrera E, et al (2013) Static, free vibration and buckling analysis of isotropic and sandwich functionally graded plates using a quasi-3D higher-order shear deformation theory and a meshless technique. *Compos B Eng* 44:657–674. https://doi.org/10.1016/j.compositesb.2012.01.089
29. Natarajan S, Manickam G (2012) Bending and vibration of functionally graded material sandwich plates using an accurate theory. *Finite Elem Anal Des* 57:32–42. https://doi.org/10.1016/j.finel.2012.03.006
30. Li Q, Iu VP, Kou KP (2008) Three-dimensional vibration analysis of functionally graded material sandwich plates. *J Sound Vib* 311:498-515. https://doi.org/10.1016/j.jsv.2007.09.018
31. Bennoun M, Houari MSA, Tounsi A (2016) A novel five-variable refined plate theory for vibration analysis of functionally graded sandwich plates. *Mech Adv Mater Struct* 23:423–431. https://doi.org/10.1080/15376494.2014.984088
32. Akay HU (1980) Dynamic large deflection analysis of plates using mixed finite elements. *Comput Struct* 11:1–11. https://doi.org/10.1016/0045-7949(80)90142-X
33. Leech JW (1965) Stability of finite difference equation for the transient response of a flat plate. *AIAA J* 3:1772–1773. https://doi.org/10.2514/3.55195
34. Yamaki N (1961) Influence of large amplitudes on flexural vibrations of elastic plates. *ZAMM: J Appl Math Mech/Z Angew Math Mech* 41:501–510. https://doi.org/10.1002/zamm.19610411204
35. Bayles DJ, Lowery RL, Boyd DE (1973) Nonlinear vibrations of rectangular plates. *ASCE J Struct Div* 99:853–864.
36. Amabili M (2008) Nonlinear vibrations and stability of shells and plates. Cambridge University Press. https://doi.org/10.1017/CBO9780511619694
37. Duc ND (2014) *Nonlinear Static and Dynamic Stability of Functionally Graded Plates and Shells*. Vietnam National University Press Hanoi.
38. Amabili M (2004) Nonlinear vibrations of rectangular plates with different boundary conditions: Theory and experiments. *Comput Struct* 82:2587–2605. https://doi.org/10.1016/j.compstruc.2004.03.077
39. Amabili M, Carra S (2009) Thermal effects on geometrically nonlinear vibrations of rectangular plates with fixed edges. *J Sound Vib* 321:936–954. https://doi.org/10.1016/j.jsv.2008.10.004
40. Amabili M (2016) Nonlinear vibrations of viscoelastic rectangular plates. *J Sound Vib* 362:142–156. https://doi.org/10.1016/j.jsv.2015.09.035
41. Duc ND, Cong PH (2013) Nonlinear postbuckling of symmetric S-FGM plates resting on elastic foundations using higher order shear deformation plate theory in thermal environments. *Compos Struct* 100:566–574. https://doi.org/10.1016/j.compstruct.2013.01.006
42. Duc ND, Tuan ND, Tran P, Quan TQ (2017) Nonlinear dynamic response and vibration of imperfect shear deformable functionally graded plates subjected to blast and thermal loads. *Mech Adv Mater Struct* 24:318–329. https://doi.org/10.1080/15376494.2016.1142024
43. Alijani F, Bakhtiari-Nejad F, Amabili M (2011) Nonlinear vibrations of FGM rectangular plates in thermal environments. *Nonlinear Dyn* 66:251–270. https://doi.org/10.1007/s11071-011-0049-8
44. Yang J, Shen HS (2002) Vibration characteristics and transient response of shear-deformable functionally graded plates in thermal environments. *J Sound Vib* 255:579–602. https://doi.org/10.1006/jsvi.2001.4161
45. Huang X-L, Shen H-S (2004) Nonlinear vibration and dynamic response of functionally graded plates in thermal environments. *Int J Solids Struct* 41:2403–2427. https://doi.org/10.1016/j.ijsolstr.2003.11.012

46. Praveen GN, Reddy JN (1998) Nonlinear transient thermoelastic analysis of functionally graded ceramic-metal plates. *Int J Solids Struct* 35:4457–4476. https://doi.org/10.1016/S0020-7683(97)00253-9

47. Upadhyay AK, Shukla KK (2013) Geometrically nonlinear static and dynamic analysis of functionally graded skew plates. *Commun Nonlinear Sci Numer Simul* 18:2252–2279. https://doi.org/10.1016/j.cnsns.2012.12.034

48. Hao YX, Zhang W, Yang J, Li SY (2011) Nonlinear dynamic response of a simply supported rectangular functionally graded material plate under the time-dependent thermalmechanical loads. *J Mech Sci Technol* 25:1637–1646. https://doi.org/10.1007/s12206-011-0501-1

49. Akbarzadeh AH, Abbasi M, Hosseini Zad SK, Eslami MR (2011) Dynamic analysis of functionally graded plates using the hybrid Fourier-Laplace transform under thermomechanical loading. *Meccanica* 46:1373–1392. https://doi.org/10.1007/s11012-010-9397-6

50. Dinh Duc N, Hong Cong P (2015) Nonlinear vibration of thick FGM plates on elastic foundation subjected to thermal and mechanical loads using the first-order shear deformation plate theory. *Cogent Eng* 2:1–17. https://doi.org/10.1080/23311916.2015.1045222

51. Jung W-Y, Han S-C, Park W-T (2016) Four-variable refined plate theory for forced-vibration analysis of sigmoid functionally graded plates on elastic foundation. *Int J Mech Sci* 111–112:73–87. https://doi.org/10.1016/j.ijmecsci.2016.03.001

52. Singh SJ, Harsha SP (2018) Nonlinear vibration analysis of sigmoid functionally graded sandwich plate with ceramic-FGM-metal layers. *J Vib Eng Technol* 8:67–84. https://doi.org/10.1007/s42417-018-0058-8

53. Singh SJ, Nataraj C, Harsha SP (2020) Nonlinear dynamic analysis of a sandwich plate with S-FGM face sheets and homogeneous core subjected to harmonic excitation. *J Sandw Struct Mater* https://doi.org/10.1177/1099636220904338

54. Singh SJ, Harsha SP (2019) Nonlinear dynamic analysis of sandwich S-FGM plate resting on pasternak foundation under thermal environment. *Eur J Mech A Solids* 76:155–179. https://doi.org/10.1016/j.euromechsol.2019.04.005

55. Wang YQ, Zu JW (2017) Large-amplitude vibration of sigmoid functionally graded thin plates with porosities. *Thin-Walled Structures* 119:911–824. https://doi.org/10.1016/j.tws.2017.08.012

56. Spriggs RM (1986) Expression for effect of porosity on elastic modulus of polycrystalline refractory materials, particularly aluminium oxide. *Sci Sinter.* 44:628–629. https://doi.org/10.1111/j.1151-2916.1961.tb11671.x

57. Pabst W, Gregorová E (2004) New relation for the porosity dependence of the effective tensile modulus of brittle materials. *J Mater Sci* 39:3501–3503. https://doi.org/10.1023/B:JMSC.0000026961.12735.2a

58. Pabst W, Tichá G, Gregorová E, Týnová E (2005) Effective elastic properties of alumina-zirconia composite ceramics part 5. Tensile modulus of alumina-zirconia composite ceramics. *Ceram Silik.*49:77–85.

59. Magnucka-Blandzi E (2008) Axi-symmetrical deflection and buckling of circular porous-cellular plate. *Thin-Walled Struct* 46:333–337. https://doi.org/10.1016/j.tws.2007.06.006

60. Kim YW (2005) Temperature dependent vibration analysis of functionally graded rectangular plates. *J Sound Vib* 284:531–549. https://doi.org/10.1016/j.jsv.2004.06.043

61. Daikh AA (2019) Temperature dependent vibration analysis of functionally graded sandwich plates resting on Winkler/Pasternak/Kerr foundation. *Mater Res Express* 6:065702. https://doi.org/10.1088/2053-1591/ab097b

62. Daikh AA, Houari MSA, Tounsi A (2019) Buckling analysis of porous FGM sandwich nanoplates due to heat conduction via nonlocal strain gradient theory. *Eng Res Express* 1:015022. https://doi.org/10.1088/2631-8695/ab38f9

63. Singh SJ, Harsha SP (2020) Thermal buckling of porous symmetric and non-symmetric sandwich plate with homogenous core and S-FGM face sheets resting on Pasternak foundation. *Int J Mech Mater Des* 16: 707–731. https://doi.org/10.1007/s10999-020-09498-7

64. Singh SJ, Harsha SP (2018) Exact solution for free vibration and Buckling of sandwich S-FGM plates on Pasternak elastic foundation with various boundary conditions. *Int J Struct Stab Dyn* 19:S0219455419500287. https://doi.org/10.1142/S0219455419500287

65. Chi SH, Chung YL (2006) Mechanical behavior of functionally graded material plates under transverse load-Part I: Analysis. *Int J Solids Struct* 43:3657–3674. https://doi.org/10.1016/j.ijsolstr.2005.04.011

66. Chung YL, Chi SH (2001) The residual stress of functionally graded materials. *J Chin Inst Civ Hydraul Eng* 13:1–9.

67. Touloukian YS (1967) *Thermophysical Properties of High Temperature Solid Materials*. Macmillan, New York.

68. Reddy J, Chin C (1998) Thermomechanical analysis of functionally graded cylinders and plates. *J Therm Stress* 37–41.

69. Joshan YS, Grover N, Singh BN (2017) A new non-polynomial four variable shear deformation theory in axiomatic formulation for hygro-thermo-mechanical analysis of laminated composite plates. *Compos Struct* 182:685–693. https://doi.org/10.1016/j.compstruct.2017.09.029

70. Dogan V (2013) Nonlinear vibration of FGM plates under random excitation. *Compos Struct* 95:366–374. https://doi.org/10.1016/j.compstruct.2012.07.024

71. Dinh Duc N, Hong Cong P (2018) Nonlinear thermo-mechanical dynamic analysis and vibration of higher order shear deformable piezoelectric functionally graded material sandwich plates resting on elastic foundations. *J Sandw Struct Mater* 20:191–218. https://doi.org/10.1177/1099636216648488

72. Duc ND, Bich DH, Cong PH (2016) Nonlinear thermal dynamic response of shear deformable FGM plates on elastic foundations. *J Therm Stress* 39:278–297. https://doi.org/10.1080/01495739.2015.1125194

73. Sundararajan N, Prakash T, Ganapathi M (2005) Nonlinear free flexural vibrations of functionally graded rectangular and skew plates under thermal environments. *Finite Elem Anal Des* 42:152–168. https://doi.org/10.1016/j.finel.2005.06.001

74. Azizian ZG, Dawe DJ (1985) Geometrically nonlinear analysis of rectangular mindlin plates using the finite strip method. *Comput Struct* 21:423–436. https://doi.org/10.1016/0045-7949(85)90119-1

10 Functionally Graded Materials

Applications and Future Challenges

Ashish Yadav, Pushkal Badoniya, Manu Srivastava,
Prashant K. Jain, and Sandeep Rathee

CONTENTS

10.1 INTRODUCTION

Functionally graded materials (FGMs) are an exquisite class of engineering materials which can either be natural or artificial. FGMs are a classic example of controlled anisotropic behavior and exhibit inhomogeneity with properties continuously varying across the dimensions normally in a defined fashion [1–3]. These materials have been comprehensively discussed in the previous chapters of this book. The different fabrication methods of FGMs such as liquid phase processing, vapor phase processing, and solid phase processing methods have already been discussed in previous

DOI: 10.1201/9781003097976-10

chapters of this book. The fabrication of FGMs using additive manufacturing techniques has also been discussed. Some case studies related to FGMs are also presented in previous chapters.

This chapter overviews and presents highlights of several applications of FGMs such as biomedical, aerospace, defense, energy, automobiles, marine, constructions, sports, optoelectronics, and so on. It then discusses the various challenges and trends in the field of FGMs and concludes with a summary to help its readers visualize this versatile engineering field of advanced materials.

10.2   APPLICATIONS FOR FGMs

In the past, the homogeneity of material properties was considered as an indicator of the efficiency of any material fabrication technique. However, as time advanced, a severe need was felt to overcome the constraints of applying homogeneous materials. Tailoring the material properties to derive FGMs was a landmark achievement in this direction. This concept basically draws its roots from nature because many natural substances like bone, bamboo, etc., offer brilliant graded properties and have inspired a lot of research on similar materials. The ever-dynamic quest of researchers for the development of exotic materials and material advancements also counts amongst the important factors in the massively keen interest of various research personnel in this direction. Figure 10.1 presents a few relevant milestones in the research and development of FGMs.

Many applications in innovative sectors like aerospace, automotive, etc., require materials to possess multi-functional properties which may or may not be conflicting. FGMs, which are a specific class of heterogeneous composite materials, can be the right solution for such requirements. The first application of FGMs was reported in a Japanese project for reducing metal ceramic interfacial thermal stress [4].

As discussed in the above section, an example of present-day FGM application of conflicting properties can be one where high surface properties such as hardness, corrosion resistance, wear resistance, etc., at the outer surface and high toughness for the bulk are required. FGMs are being increasingly used by many industries and provide better promise in applications like thermoelectric generators, rocket heat

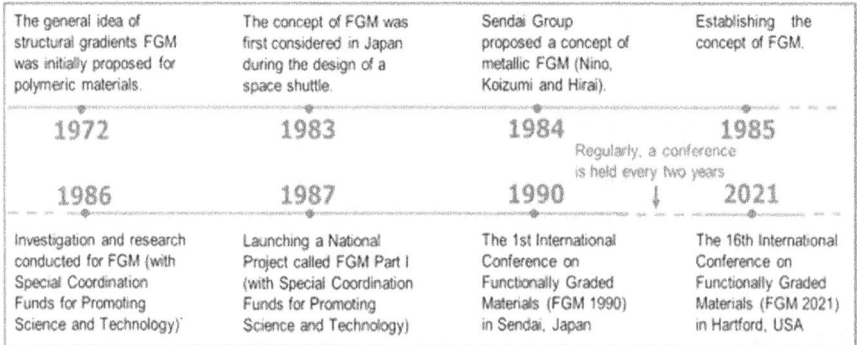

The general idea of structural gradients FGM was initially proposed for polymeric materials.	The concept of FGM was first considered in Japan during the design of a space shuttle.	Sendai Group proposed a concept of metallic FGM (Nino, Koizumi and Hirai).	Establishing the concept of FGM.
1972	**1983**	**1984**	**1985**
		Regularly, a conference is held every two years	
1986	**1987**	**1990**	**2021**
Investigation and research conducted for FGM (with Special Coordination Funds for Promoting Science and Technology)	Launching a National Project called FGM Part I (with Special Coordination Funds for Promoting Science and Technology)	The 1st International Conference on Functionally Graded Materials (FGM 1990) in Sendai, Japan	The 16th International Conference on Functionally Graded Materials (FGM 2021) in Hartford, USA

FIGURE 10.1   Relevant milestones in FGM R&D [5].

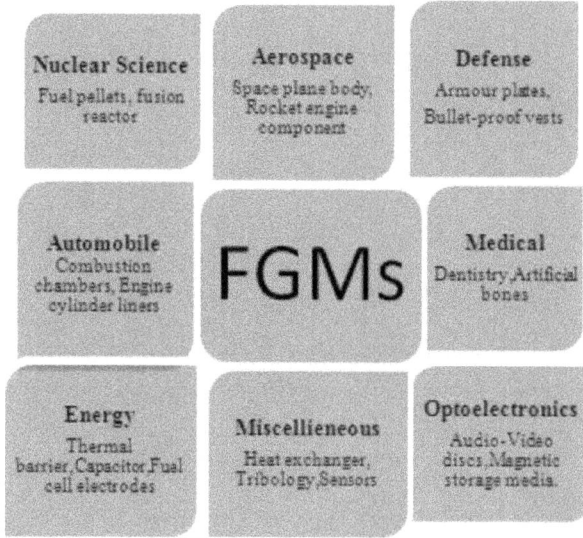

FIGURE 10.2   Application areas of FGMs [6].

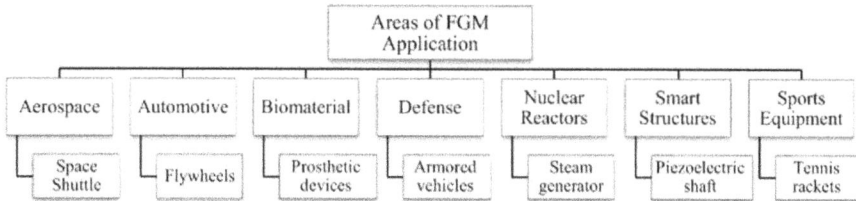

FIGURE 10.3   Application areas and some examples of FGMs [7].

shields, heat engine components, heat exchangers, etc. Different types of FGM applications such as in the biomedical, aerospace, defense, electrical/electronics, marine, optoelectronics, sport, energy, automobile, and other industries have been presented non-exhaustively in Figures 10.2 and 10.3. An attempt to briefly highlight the various main FGM applications is undertaken in the following sections.

10.2.1 BIOMEDICAL APPLICATIONS

Medicinal components must exhibit non-toxicity, durability, thermal conductivity, strength, biocompatibility, corrosion resistance, and abrasion resistance that cannot be achieved with a single composition with a uniform structure [8]. The human body has many tissues. These human tissues may at times suffer from damage which cannot be treated, thereby requiring mandatory replacement by alternative parts. With the help of FGMs, we can replace the damaged tissues such as bones and teeth as shown in Figure 10.4 [8].

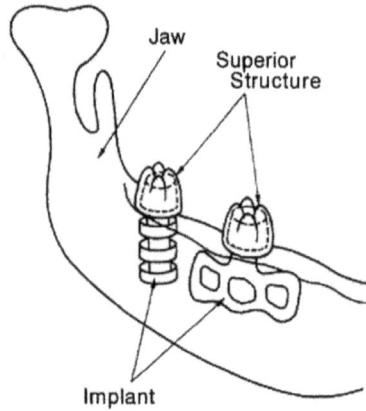

FIGURE 10.4   Functionally graded material dental implant [9].

Recently, TiN/HA, Ti/SiO$_2$, Ti/Co, and Ti/ZrO$_2$, different types of ceramic composite FGMs, have been investigated for biomaterials and bio-ceramics. In the present medical scenario, knee, dental implants, and prosthetic parts practice deals with replacing and repairing synovium joint (bones that move each other), missing tooth structure, etc. FGMs can be used in all these applications. They are also used to repair cartilages (resilient and smooth elastic tissue) [10, 11]. One such case study for bone modeling using a fused filament fabrication AM technique has also been included in the present book as Chapter 6.

10.2.2   AEROSPACE APPLICATIONS

FGMs were initially used in spacecraft for decreasing temperature stress from the inner to the outer surface. However, FGMs are used for various aerospace applications in the present times [12]. Now, most aircraft components such as reflectors, solar panels, turbine wheels, rocket engines, etc., use FGMs as shown in Figure 10.5 [14]. All these components of spacecraft are designed for higher temperatures at a particular area, a requirement which can be easily fulfilled by FGMs. FGMs such as Al$_2$O$_3$/ Al alloy with good temperature and erosion resistance are used to make aviation engine parts and bearing lines. Zhao et al. [15] developed a different approach for a tungsten and tungsten/copper FG composite by introducing the hot pressing approach to upgrade properties for applications of aircraft. Kumar et al. [12] prepared polymer-ceramic continuous fiber-reinforced FG composites for aerospace applications.

10.2.3   DEFENSE APPLICATIONS

Currently in defense, the main objective is to reduce the weight of weapons, armors, and tanks, and inhibit the distribution of cracks, etc. Most of these requirements can be fulfilled by using FGMs. A pierce-resistant material for the production of

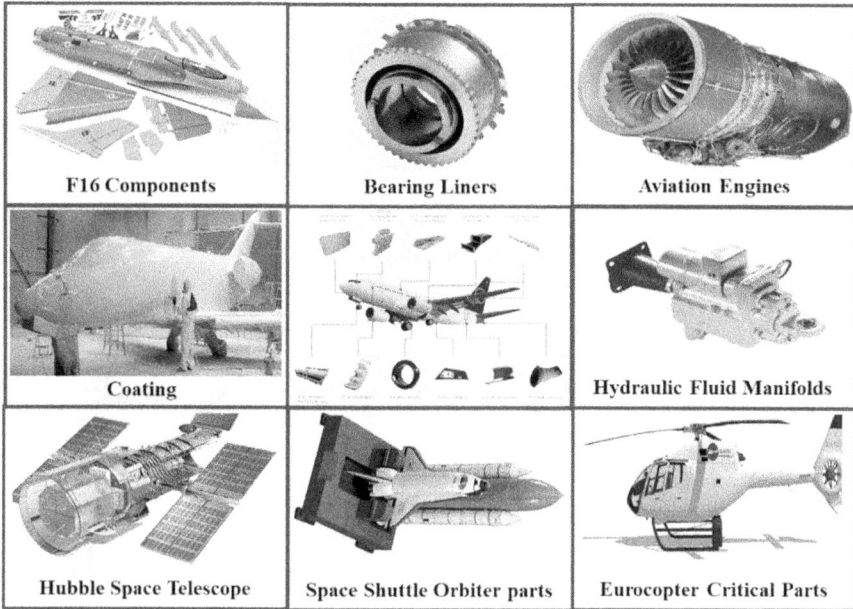

FIGURE 10.5   FGM parts in aerospace applications [13].

bulletproof jackets, firing pins, rods, tanks, and axles can be developed by using FGMs [16, 17]. Another important application of FGMs is the body of bullet-proof vehicles.

Huang et al. [18] used FGM principles to produce armor with higher impact resistance and less weight from Al_2O_3/ZrO_2 ceramic composites. After verifying the results from simulation and experiment they concluded that FGM Al_2O_3/ZrO_2 ceramic composite had good properties compared to conventional composites.

10.2.4   ENERGY APPLICATIONS

FGMs are utilized in energy generation owing to their ability to be used in many parts of energy devices, for example solar energy device parts, coating of the blades of turbines in power plants, sensors, batteries, etc. [19, 20]. FGM structures have high energy potential. The concept of FGMs is now applicable to many hybrid systems in addition to systems operating on a single material.

10.2.5   AUTOMOBILE INDUSTRY APPLICATIONS

Components in the automobile industry like shock absorbers, driveshafts, brakes crankshafts, and governors, as shown in Figure 10.6, are required to be designed under temperature, high and low compulsions. Therefore, FGMs are very suitable for these components. Also the possibility of FG coating on the automotive body is

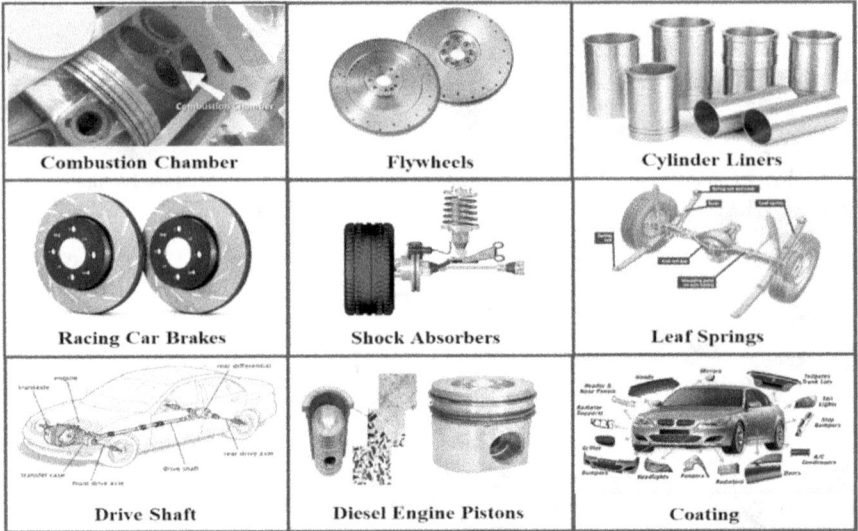

FIGURE 10.6 FGM parts in the automobile industry [13].

an extremely lucrative option owing to the high cost of FGMs which limits their full-scale use in the automobile industry [21].

Kimberly R. [22] prepared FG Al_2O_3/Al_2TiO_5 composite ceramics for microstructure, thermal, and physical properties analysis for automobile applications, and results show that the FGMs upgraded thermal and mechanical qualities for the car brake system. Ram et al. [23] developed FG Al/Mg_2Si composites for elevated temperature tensile properties for automotive cylinder liners, and the results reflected an upgraded microstructure and mechanical effect for liners.

10.2.6 MARINE APPLICATIONS

FGMs are also used in marine and submarine applications because of their superior hardness, toughness, structural, and manufacturing flexibility features. Owing to their graded properties, FGMs find extensive use in propeller shafts, sonar domes, cylinders, pressure hulls, sensors, propulsion shafts, machinery, and auxiliary machining, and so on.

10.2.7 CONSTRUCTION APPLICATIONS

In the construction field, there are many challenges like material availability according to the building structure, cost of material, strength, environmental impact, and so on. By using FGMs we can fulfill all these demands, i.e. structure and varying material composition [24].

10.2.8 OPTO-ELECTRONICS APPLICATIONS

FGMs are used in opto-electronic devices because of the refractive index gradients that can be achieved. Examples include lasers, magnetic storage media, sensors, fibers, solar cells, computer circuit boards, and semiconductor applications.

10.2.9 MACHINES/EQUIPMENT APPLICATIONS

FGMs find extensive utility in fabricating several engineering components like cutting tool material, molds for forming, engine blocks, etc., since there is considerable improvement in strength, temperature, and wear as well as corrosion resistance of parts by the use of FGMs [25]. Recent years have witnessed the utilization of FGM-based WC/Co and TiCN–WC/Co as cutting tool materials, for instance in turning, grinding and milling, twist drilling, etc., for limiting wear at surfaces as well as enhancing hardness which results in increased tool survival rates, feed rates, and machining speed as well as reducing machining time [26]. The design and fabrication of FG WC–Co ceramic composite materials using liquid phase sintering for estimating the weight-fraction and particle dimension effects of Co gradient upon microstructural as well as mechanical characteristics for utilization as cutting tools has been achieved by Li et al. [27]. They concluded that the gradation of particles and thus the desired properties vary with capillary force and weight fraction. Zhao et al. [28] fabricated FG Al_2O_3–TiC and Al_2O_3–(W, Ti) C tool material for evaluating quenching as well as the bending effect upon thermal shock resistance. Xing et al. [29] reported similar findings for FG Al_2O_3/TiC composites fabricated via the powder metallurgy technique. In almost every application, it was found that the performance of FG-based tools/machine parts was either comparable or surpassed the quality of conventionally utilized material.

10.2.10 SPORTS APPLICATIONS

FGMs, especially those fabricated via AM techniques, are utilized for sports equipment like golf clubs, skis, tennis rackets, etc. This is chiefly attributed to the graded characteristics obtained and thus reduced weight and friction, enhanced durability, higher strength to weight ratio, hardness, wear resistance, and several other desirable attributes.

10.2.11 MISCELLANEOUS APPLICATIONS

Apart from those mentioned above, there are several other areas where FGMs are being increasingly used. These include medicine, medical equipment, safety equipment, nuclear reactor components, turbine blades, heat exchangers, sensors, advanced materials, and so on [30]. FGM's application spectrum is expected to increase further, subject to reduction in the fabrication costs.

The various specific applications along with broad areas and FGM systems utilized are presented in Table 10.1 [13].

TABLE 10.1

Some Examples of FGM Applications in Different Sectors [13]

Main Applications	Examples	FGM Systems	Ref.
Aerospace applications	Rocket nozzles, drive shafts, rings, wings, rotor drive shafts, rotary launchers, landing gear doors, spacecraft truss structures, telescope metering truss assembly, structure blades, and solar panels	$Al–Al_2O_3$, SiC–C, C–C, WC–Co, Be–Al, TiAl–SiC, Al_2O_3/ZrO_2, Al_2O_3–lanthanum hexaaluminate (LHA)	[31–35]
Automotive applications	Brake rotors, diesel engine pistons, cylinder liners, pulleys, shock absorbers, flywheels, leaf springs, drive shafts, combustion chambers, CNG storage cylinders, and racing car brakes	$Al_2O_3–Al_2TiO_5$, Al–SiC, $Al–Al_2O_3$, SiC–SiC, Al–C, E-glass-epoxy, and $Al–Mg_2Si$	[21–23, 36]
Machinery and equipment applications	Cutting tool inserts, wind turbine blades, pressure vessels, drilling motor shafts, fuel tanks, machine parts, laptop cases, firefighting air bottles, eyeglass frames, X-ray tables, MRI scanner cryogenic tubes, racing bicycle frames, golf clubs, tennis rackets, skis, musical instruments, and racing vehicle frames	$Cu–ZrO_2$/ $Cu–C–MoS_2–ZrO_2$, $TiCN–WC$/Co, WC/Co	[28, 29, 37–39]
Biomedical applications	Artificial skin, bone implants, dental implants, bone cartilage repair	Ti/HA, Al_2O_3/ZrO_2, $Al_2O_3–SiC–ZrO_2$, $HA–Al_2O_3–ZrO_2$, SS316L-HA, ZrO_2/porcelain	[10, 40–46]
Defense applications	Armor plates and bulletproof vests, guide rods, precision rollers, shafts, tubes, latches, axles housings, and firing pins	TiB_2–Ti, Al–SiC, $ZrO–Zr_2CN/Si3N4$, $Al_2O_3–ZrO_2$	[16, 18, 47, 48]
Energy applications	Cathode materials, evaporator tubes, thermal power generators, solar power components and energy conversion devices, capacitors, sensors, electrodes, and turbine blades	$Al–Mg_2Si$, Ni–ZrO_2	[49, 50]
Electronic/optoelectronic applications	Fuel cells, vacuum switch tubes, electrical contact materials, heat sink, integrated circuits, semiconductor devices, and electronic substrates	Ni–ZrO_2, Ni–W/ZrO_2, Mo–Cu, W–Cu	[19, 20, 51–53]

10.3 FUTURE TRENDS

FGMs qualify as perfect candidates to solve the issue of increased demand for parts that require conflicting properties within their own selves. This explains the varied and vast spectrum of applications of FGMs with tailored properties. Owing to the advantages and unique properties offered by this exquisite class of materials, there is no denying the fact that FGMs have a huge future scope.

A noteworthy point here is that even though there has been appreciable improvement in the development and utilization of FGMs in recent years, many aspects need to be fully explored to utilize the full potential of these materials. Figure 10.7 shows a few domains that need to be thoroughly explored.

By virtue of their being designed and fabricated for specific customer requirements, FGMs possess high potential to influence and modify the utilization of engineering materials in the near future. They demonstrate huge capabilities for supporting our innovative technologies globally. Areas with high potential would include those where typical gradients of strength, toughness, hardness, etc., will be required, biomedical applications, bio-inspired technologies, etc. Apart from this, research avenues will be open for expansion if the cost of these materials can be kept minimal, either by the development of special fabrication techniques or by some other routes. The transition of FGM technology into the fourth industrial revolution would necessitate undertaking rigorous research initiatives to develop solutions of different avenues. Clearly defined FGM applications along with property specifications to take correct decisions on functionality, the right mix of input materials as well as fabrication method would go a long way in the full utilization of this concept while ensuring efficient and cost-effective practices.

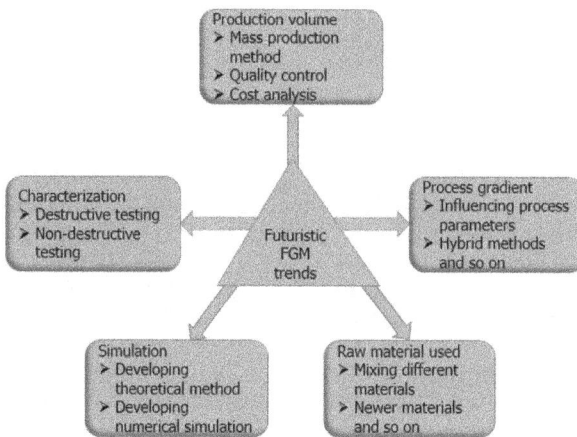

FIGURE 10.7 Probable FGM future trends.

10.4 SUMMARY AND CONCLUDING REMARKS

FGMs have superior hardness, toughness, and structural and manufacturing flex-ibility features owing to which they find use in defense, biomedical, automobiles, aerospace, energy sector, sports, etc. In recent years, many researchers have accom-plished appreciable research in the field of FGM development and improvement. However, in-depth investigations in many areas are still needed.

FGMs are materials with tailored properties that vary with predefined gradient and can be developed by a variety of processing techniques. Each technique has its own applications, advantages, and disadvantages. The major research gaps in this field are: optimizing the size of particles, stress analysis, subatomic particles in FGMs, understanding the mechanism, analysis of the mechanism, process param-eter optimization, virtual and physical simulation, characterization, enhancing mate-rial properties, void formation, newer materials, the use of multiple feeding systems, multi-material parts, choice of weight and volume fraction, and so on.

The future demands of FGM will specially be those in which extraordinarily spe-cific combinations of mechanical, thermal, and chemical characteristics are desir-able. An appreciable landmark in the FGM area will be containing the high cost of FGM production so low-cost techniques for large parts and complex shapes still have to be studied.

This edited book presents a compilation of research ideas by different contribu-tors in the field of FGM. The current work can serve as a milestone for aspiring researchers and material scientists who have a keen interest in the field of material advancements, especially FGMs.

REFERENCES

1. Srivastava, M., Rathee, S., Maheshwari, S., Kundra, T.K., Design and process-ing of functionally graded material: Review and current status of research. In 7th International Conference on 3D Printing & Additive Manufacturing Technologies- AM 2017, GLOBAL SUMMIT 2017: Bengaluru, India.
2. Srivastava, M., Maheshwari, S., Kundra, T.K., Rathee, S., Yashaswi, R., Sharma, S.K., Virtual design, modelling and analysis of functionally graded materials by fused deposition modeling. *Materials Today: Proceedings*, 2016. **3**(10, Part B): p. 3660–3665.
3. Srivastava, M., Maheshwari, S., Kundra, T.K., Virtual modelling and simulation of functionally graded material component using FDM technique. *Materials Today: Proceedings*, 2015. **2**(4): p. 3471–3480.
4. Nejad, M.Z., Alamzadeh, N., Hadi, A., Thermoelastoplastic analysis of FGM rotating thick cylindrical pressure vessels in linear elastic-fully plastic condition. *Composites Part B: Engineering*, 2018. **154**: p. 410–422.
5. Saleh, B., Jiang, J., Ma, A., Song, D., Yang, D., Xu, Q., Review on the influence of differ-ent reinforcements on the microstructure and wear behavior of functionally graded alu-minum matrix composites by centrifugal casting. *Metals and Materials International*, 2020. **26**(7): p. 933–960.
6. Gupta, A., Talha, M., Recent development in modeling and analysis of functionally graded materials and structures. *Progress in Aerospace Sciences*, 2015. **79**: p. 1–14.

7. El-Galy, I.M., Saleh, B.I., Ahmed, M.H., Functionally graded materials classifications and development trends from industrial point of view. *SN Applied Sciences*, 2019. **1**(11): p. 1378.

8. Saleh, B., Jiang, J., Ma, A., Song, D., Yang, D., Effect of main parameters on the mechanical and wear behaviour of functionally graded materials by centrifugal casting: A review. *Metals and Materials International*, 2019. **25**(6): p. 1395–1409.

9. Watari, F., Yokoyama, A., Saso, F., Uo, M., Kawasaki, T., Fabrication and properties of functionally graded dental implant. *Composites Part B Engineering*, 1997. **28**(1): p. 5–11.

10. Pompe, W., Worch, H., Epple, M., Friess, W., Gelinsky, M., Greil, P., Hempel, U., Scharnweber, D., Schulte, K., Functionally graded materials for biomedical applications. *Materials Science and Engineering: Part A*, 2003. **362**(1): p. 40–60.

11. El-Galy, I.M., Bassiouny, B.I., Ahmed, M.H., Empirical model for dry sliding wear behaviour of centrifugally cast functionally graded Al/SiCp composite. *Key Engineering Materials*, 2018. **786**: p. 276–285.

12. Kumar, S., Murthy Reddy, K.V.V.S., Kumar, A., Rohini Devi, G., Development and characterization of polymer–ceramic continuous fiber reinforced functionally graded composites for aerospace application. *Aerospace Science and Technology*, 2013. **26**(1): p. 185–191.

13. Saleh, B., Jiang, J., Fathi, R., Al-hababi, T., Xu, Q., Wang, L., Song, D., Ma, A., 30 years of functionally graded materials: An overview of manufacturing methods, applications and future challenges. *Composites Part B: Engineering*, 2020. **201**: p. 108376.

14. Udupa, G., Rao, S.S., Gangadharan, K.V., Functionally graded composite materials: An overview. *Procedia Materials Science*, 2014. **5**: p. 1291–1299.

15. Zhao, P., Wang, S., Guo, S., Chen, Y., Ling, Y., Li, J., Bonding W and W–Cu composite with an amorphous W–Fe coated copper foil through hot pressing method. *Materials & Design*, 2012. **42**: p. 21–24.

16. Chin, E.S.C., Army focused research team on functionally graded armor composites. *Materials Science and Engineering: Part A*, 1999. **259**(2): p. 155–161.

17. Chen, W.W., Rajendran, A.M., Song, B., Nie, X., Dynamic fracture of ceramics in armor applications. *Journal of the American Ceramic Society*, 2007. **90**(4): p. 1005–1018.

18. Huang, C.-Y., Chen, Y.-L., Design and impact resistant analysis of functionally graded Al2O3–ZrO2 ceramic composite. *Materials & Design*, 2016. **91**: p. 294–305.

19. Müller, E., Drašar, Č., Schilz, J., Kaysser, W.A., Functionally graded materials for sensor and energy applications. *Materials Science and Engineering: Part A*, 2003. **362**(1): p. 17–39.

20. Niino, M., Kisara, K., Mori, M., Feasibility study of FGM technology in space solar power systems (SSPS). *Materials Science Forum*, 2005. **492–493**: p. 163–170.

21. Udupa, G., Shrikantha Rao, S., Gangadharan, K.V. Future applications of carbon nanotube reinforced functionally graded composite materials. In IEEE-International Conference On Advances In Engineering, Science And Management (ICAESM -2012). 2012.

22. Kimberly, R., Oo, Z., Sujan, D., Microstructure analysis, physical and thermal properties of Al2O3–Al2TiO5 functionally graded ceramics for the application of car brake rot. *Pertanika Journal Science and Technology*, 2015. **23**(1): p. 153–161.

23. Ram, S.C., Chattopadhyay, K., Chakrabarty, I., High temperature tensile properties of centrifugally cast in-situ Al-Mg2Si functionally graded composites for automotive cylinder block liners. *Journal of Alloys and Compounds*, 2017. **724**: p. 84–97.

24. Craveiro, F., Matos, J., Bártolo, H., Bártolo, P., *Automation for Building Manufacturing*, 2011. p. 451–455.

25. Lengauer, W., Dreyer, K., Functionally graded hardmetals. *Journal of Alloys and Compounds*, 2002. **338**(1): p. 194–212.

26. Konyashin, I., Zaitsev, A.A., Sidorenko, D., Levashov, E.A., Konischev, S.N., Sorokin, M., Hlawatschek, S., Ries, B., Mazilkin, A.A., Lauterbach, S., Kleebe, H.J., On the mechanism of obtaining functionally graded hardmetals. *Materials Letters*, 2017. **186**: p. 142–145.

27. Li, L., Li, Y., Development and trend of ceramic cutting tools from the perspective of mechanical processing. *IOP Conference Series: Earth and Environmental Science*, 2017. **94**: p. 012062.

28. Zhao, J., Ai, X., Deng, J., Wang J., Thermal shock behaviors of functionally graded ceramic tool materials. *Journal of the European Ceramic Society*, 2004. **24**(5): p. 847–854.

29. Xing, A., Jun, Z., Chuanzhen, H., Jianhua, Z., Development of an advanced ceramic tool material: Functionally gradient cutting ceramics. *Materials Science and Engineering: Part A*, 1998. **248**(1): p. 125–131.

30. Malinina, M., Sammi, T., Gasik, M.M., Corrosion resistance of homogeneous and FGM coatings. *Materials Science Forum*, 2005. **492–493**: p. 305–310.

31. Kim, J.I., Kim, W.J., Choi, D.J., Park, J.Y., Ryu, W.S., Design of a C/SiC functionally graded coating for the oxidation protection of C/C composites. *Carbon*, 2005. **43**(8): p. 1749–1757.

32. Gururaja, U., ShrikanthaRao, S., Gangadharan, K.V., *A Review of Carbon Nanotube Reinforced Aluminium Composite and Functionally Graded Composites as a Future Material for Aerospace*. 2014.

33. Limarga, A.M., Widjaja, S., Yip, T.H., Mechanical properties and oxidation resistance of plasma-sprayed multilayered Al2O3/ZrO2 thermal barrier coatings. *Surface and Coatings Technology*, 2005. **197**(1): p. 93–102.

34. Negahdari, Z., Willert-Porada, M., Scherm, F., Development of Novel Functionally Graded Al2O3-Lanthanum Hexaaluminate Ceramics for Thermal Barrier Coatings. *Materials Science Forum*, 2010. **631–632**: p. 97–102.

35. Zhang, L.X., Zhang, B., Sun, Z., Pan, X.Y., Shi, J.M., Feng, J.C., Preparation of graded double-layer materials for brazing C/C composite and TC4. *Journal of Alloys and Compounds*, 2020. **823**: p. 153639.

36. Li, Y., Jian, S., Min, Z., Application of ceramics metal functionally graded materials on green automobiles. *Key Engineering Materials*, 2004. **280–283**: p. 1925–0.

37. Bahri, A., Salehi, M., Akhlaghi, M., Using a pseudo-functionally graded interlayer in order to improve the static and dynamic behavior of wind turbine blade T-bolt root joints. *Composite Interfaces*, 2014. **21**(8): p. 749–770.

38. Cho, J.R., Park, H.J., High strength FGM cutting tools: Finite element analysis on thermoelastic characteristics. *Journal of Materials Processing Technology*, 2002. **130–131**: p. 351–356.

39. Xu, F., Zhang, X., Zhang, H., A review on functionally graded structures and materials for energy absorption. *Engineering Structures*, 2018. **171**: p. 309–325.

40. Akmal, M., Khalid, F.A., Hussain, M.A., Interfacial diffusion reaction and mechanical characterization of 316L stainless steel-hydroxyapatite functionally graded materials for joint prostheses. *Ceramics International*, 2015. **41**(10, Part B): p. 14458–14467.

41. Miao, X., Sun, D., Graded/gradient porous biomaterials. *Materials*, 2010. **3**(1): p. 26–47.

42. Petit, C., Montanaro, L., Palmero, P., Functionally graded ceramics for biomedical application: Concept, manufacturing, and properties. *International Journal of Applied Ceramic Technology*, 2018. **15**(4): p. 820–840.

43. Gupta, B., Kakkar, K., Gupta, L., Gupta, A., Anesthetic considerations for a parturient with pulmonary hypertension. *Indian Anaesthetists Forum*, 2017. **18**: p. 39–45.

44. Bakar, W., Basri, S.N., Jamaludin, S.N., Sajjad, A., Functionally graded Materials: An overview of dental applications. *World Journal of Dentistry*, 2019. **9**: p. 137–144.

45. Tharaknath, S, Ramkumar, R., Lokesh, B., Design and analysis of hip prosthesis using functionally graded Material. *Middle East Journal of Scientific Research*, 2016. **24**: p. 124–132.
46. Almasi, D., Sadeghi, M., Lau, W.J., Roozbahani, F., Iqbal, N., Functionally graded polymeric materials: A brif review of current fabrication methods and introduction of a novel fabrication method. *Materials Science and Engineering. Part C*, 2016. **64**: p. 102–107.
47. Gupta, N., Prasad, V.V.B., Madhu, V., Basu, B., Ballistic studies on TiB2-Ti functionally graded armor ceramics. *Defence Science Journal*, 2012. **62**: p. 382–389.
48. Li, L., Cheng, L., Fan, S., Gao, X., Xie, Y.P., Zhang, L., Fabrication and dynamic compressive response of laminated ZrO–Zr2CN/Si3N4 ceramics. *Ceramics International*, 2015. **41**(7): p. 8584–8591.
49. Woolley, R.J., Skinner, S.J., Functionally graded composite La2NiO4+δ and La4Ni3O10−δ solid oxide fuel cell cathodes. *Solid State Ionics*, 2014. **255**: p. 1–5.
50. Vaidya, R., *Plasma Sprayed Functionally Graded and Layered MoSi2-A1203 Composites for High Temperature Sensor Sheath Application*, 2001, Los Alamos National Lab., Los Alamos, NM.
51. Bharti, I, Gupta, N., Gupta, K.M., Novel applications of functionally graded nano, optoelectronic and thermoelectric materials. *International Journal of Materials Mechanics and Manufacturing* 2013. **1**(3): p. 221–224.
52. Li, S.-B., Xie, J.-X., Processing and microstructure of functionally graded W/Cu composites fabricated by multi-billet extrusion using mechanically alloyed powders. *Composites Science and Technology*, 2006. **66**(13): p. 2329–2336.
53. Wośko, M., Paszkiewicz, B., Piasecki, T., Szyszka, A., Paszkiewicz, R., Tłaczała, M., Applications of functionally graded materials in optoelectronic devices. *Optica Applicata*, 2005. **35**: p. 663–667.

Index

For Product Safety Concerns and Information please contact our EU
representative GPSR@taylorandfrancis.com
Taylor & Francis Verlag GmbH, Kaufingerstraße 24, 80331 München, Germany

* 9 7 8 0 3 6 7 5 6 4 8 5 8 *